PHY OF
water

THE BROWN
FOUNDATIONS OF GEOGRAPHY
SERIES

Consulting Editor
ROBERT H. FUSON
University of South Florida

A GEOGRAPHY OF

Agriculture
James R. Anderson, University of Florida
Transportation and Business Logistics
Edwin J. Becht, University of Oklahoma
Plants and Animals
David J. de Laubenfels, Syracuse University
Geography
Robert H. Fuson, University of South Florida
The Atmosphere
John J. Hidore, Indiana University
Population and Settlement
Maurice E. McGaugh, Central Michigan University
Industrial Location
E. Willard Miller, The Pennsylvania State University
Water
Ralph E. Olson, University of Oklahoma
Earth Form
Stuart C. Rothwell, University of South Florida
Minerals
Walter H. Voskuil, University of Nevada

THE BROWN
FOUNDATIONS OF GEOGRAPHY
SERIES

A GEOGRAPHY OF

water

RALPH E. OLSON
University of Oklahoma

WM. C. BROWN COMPANY PUBLISHERS
DUBUQUE, IOWA

THE BROWN
FOUNDATIONS OF GEOGRAPHY
SERIES

Consulting Editor
ROBERT H. FUSON
University of South Florida

ISBN 0–697–05159–5

Library of Congress Catalog Card Number: 77-118883

Printed in the United States of America.

Geography is one of man's oldest sciences, yet it is as new as the Space Age. Knowledge of the earth obtained from satellite photography and measurement, remote sensing of the environment, and by means of other sophisticated techniques are really but a stage in the evolutionary process that began with ancient man's curiosity about his surroundings. Man has always been interested in the earth and the things on it. Today this interest may be channeled through the discipline of geography, which offers one means of organizing a vast amount of physical and cultural information.

The **Brown Foundations of Geography Series** has been created to facilitate the study of physical, cultural, and methodological geography at the college level. The **Series** is a carefully selected group of titles that covers the wide spectrum of basic geography. While the individual titles are self-contained, collectively they comprise a modern synthesis of major geographical principles. The underlying theme of each book is to foster an awareness of geography as an imaginative, evolving science.

To my wife, Margaret, and my three children, Karen, Susan, and Gordon, with whom I share many happy memories of travel by land and water.

Preface

The story of water is the story of a natural earth substance so widely available and in such universal use that most people take it for granted. Its physical and chemical behavior, however, is so complex that scientists still feel they have much to learn about it. Its extensive presence on earth and its complete or near absence on other planets suggest astronomical mysteries worthy of careful investigation.

Since geography is the study of "differences from place to place"—to use the simple but perceptive phrase of Preston James—there is a geography of water. It consists of a description and analysis of the variable occurrence of water over the earth, its behavior as a natural substance in a broad range of environmental circumstances, and its role in the life and activity of man. In discussing water from a primarily geographic point of view, the author has occasionally summarized in a few paragraphs what other students of water (geologists, hydrologists, economists, conservationists) have written whole books about.

Writing a geography of water is, like so much geographic writing, a task of synthesis and organization. It is a challenge in interpretation. Geographic writing often selects a single element or aspect of the physical or human environment as a focus for study. In this book the focus selected is water, an environmental factor so important and pervasive that without it the earth's physical environment would be unrecognizable and the biological-social environment nonexistent.

Ralph E. Olson

Contents

A Multiple-Purpose River Basin Development

A Multi-Purpose River Basin Development. This imaginative sketch shows some of the many ways in which water is trapped, processed, and distributed in the service of man. From **A Water Policy for the American People,** vol. I, opposite p. 1, courtesy President's Water Resources Policy Commission.

Chapter 1

An Introduction to Water

The Importance of Water

When the ancient Greek scholar, Thales of Miletus, in the early part of the sixth century B.C. declared water to be the primordial substance, the fundamental stuff from which all other materials on earth derive and to which they eventually return, he was perhaps engaging in philosophy rather than in science.[1] Water was extremely important to the early Greeks, as it is to the Greeks of the present time. Systematic efforts to understand the nature of the physical universe, and the opportunities for man afforded by it, have been more or less continuous since Thales' time, increasing in intensity and presumably also in accuracy as we approach the present. Knowledge of water is now impressively vast, but it is still far from complete.

Such fields as *oceanography, meteorology, glaciology, limnology, hydrology, water engineering, water economics,* and *water conservation* present many tantalizing questions for further research. At the same time as the research frontiers are pursued, there is periodic need for a fresh summary of the present state of knowledge in any given field. This book is an effort to organize and present in concise form those aspects of water which are significant in physical and human geography.

Water (which in its totality may be called the *hydrosphere*) is so much a part of daily life that people are scarcely aware of the many ways in which they encounter it and depend upon it. Without it, organic life as it is known, could not exist. Both plants and animals incorporate large amounts of it in their cellular structure and suffer serious consequences when deprived of it. Man can survive with little or no food for up to several weeks, but without water he can expect to live

[1]Herbert N. Couch, *Classical Civilization: Greece,* Englewood Cliffs, N. J.: Prentice-Hall, Inc., 1940, pp. 157-159.

no more than two or three days. It is perhaps not surprising that whole milk is about 87 percent water, but even much of the so-called solid food that man eats consists largely of water. Ordinary white bread, made with wheat flour, is more than one-third water, and fresh corn muffins are almost one-half. Raw beefsteak is about 63 percent water; potatoes, 78 percent water; and tomatoes close to 90 percent water.[2] The human body itself is about 71 percent water; and to maintain a proper liquid level for carrying on such essential bodily functions as breathing, digestion of food, body metabolism, blood circulation, glandular activity, temperature control, perspiration, and waste excretion, the average person living in a temperate climate with moderate activity must consume from five to six pints of water per day. Somewhat more than one-half of this water is normally consumed in the form of fluids (water, milk, or other beverages) while the remainder is taken in with the rest of the diet and transformed as needed in the body by the oxidation of sugars, starches, and fats.[3]

In primitive societies, per capita water consumption is only about one to five gallons per day for all purposes including direct consumption, cooking, and sanitation. Modern urban man in western societies, however, has learned to use, and has come to expect, several times as much. As early as 1920 the average per capita use of water in American cities was 115 gallons per day, and by 1960 it had increased to 150 gallons per day. Of this 150 gallons of domestic-use water, which people expect to be made available by public or private water systems, public uses (fire extinguishing, street cleaning, park maintenance, public building supply, and the like) demand about 10 gallons, commerce and industry about 70 gallons, and private domestic use about 70 gallons. Car washing, air conditioning, and lawn and garden sprinkling are somewhat seasonal uses of domestic water, but other major domestic uses continue throughout the year with great regularity. It takes about three gallons of water to flush a standard-type toilet, 20 to 40 gallons for a shower or tub bath, 10 gallons to wash a load of dishes in a mechanical dishwater, and 20 to 30 gallons for a load of clothes in a standard-type washing machine.[4]

Of the 270 billion gallons per day of water taken from streams, lakes, and ground water sources in the United States in 1960, as estimated by the U.S. Geological Survey, 51 percent was attributed to industry, six percent to public water supplies, one percent to rural domestic uses,

---

[2]Thomson King, *Water: Miracle of Nature,* New York: Collier Books, 1961, p. 19.

[3]United States Department of Agriculture, *Water: The Yearbook of Agriculture, 1955,* Washington: Government Printing Office, 1955, pp. 2-3.

[4]Luna B. Leipold and Walter B. Langbein, *A Primer on Water,* United States Geological Survey, Miscellaneous Reports, Washington: Government Printing Office, 1960, pp. 32-33.

and 40 percent to irrigation. Since much of the water withdrawn for general industrial and domestic use is returned directly to streams and lakes and so may be used again, it need not be considered "consumptive." Approximately 85 percent of the consumptive use of water is in irrigation and is concentrated largely in the western half of the country.

The dependence of people on the availability of water is evident at many different levels. Without the water vapor in the air, human skin would crack or harden and the lungs would shrivel painfully. Without water vapor in the air and the possibility of chilling it to form rain or snow, the surface of the land would be a barren desert. The vast ocean reservoir, which covers about 71 percent of the globe, is the ultimate source for most of the water that is present in the air or found on or under the land surface. Man looks to the water for protection, for subsistence, for industrial energy, for processing materials, for low-cost transportation, and for much of his recreation. He would be helpless—indeed, he would not even be here—without it.

Amount and General Distribution of Water

While the ocean water covers most of the earth's surface, and at times in the geologic past covered even more of the surface than it does at present, it is in reality a very thin layer, averaging less than two and a half miles thick and reaching a maximum depth of only a little more than seven miles. The total water supply has remained approximately constant during the whole of earth history, but its distribution and its physical and chemical properties have changed somewhat with the tectonic movements of the crust, with the formation of sedimentary and other porous rocks, with the addition of soluble materials from the land, and with temperature and other climatic conditions.

Of the total world water supply of 326,071,300 cubic miles, about 317,000,000 cubic miles, or 97.2 percent of it, is in the world's oceans. Most of the remaining 2.8 percent is locked up in icecaps and glaciers, mainly in Antarctica and Greenland, and the balance, between 0.6 and 0.7 percent of the total supply, is the water on and beneath the land areas.

Water has become a subject of great popular interest. In late 1965 one of the popular American news magazines published a rather striking diagrammatic sketch of the world water cycle, showing the approximate disposition of the global water supply and the daily rate of turnover in the hydrologic cycle.[5] Most of the fresh (*i.e.*, non-saline) water

[5] "Water: Worldwide Use and Misuse," *Time*, vol. 86, no. 14, October 1, 1965, pp. 70-79B.

is beneath the surface and is commonly referred to as ground water. Some of this ground water, about half of which is present within a half mile of the surface, is too mineralized for much economic use. Most of the 1,000,000 or so cubic miles of ground water within half a mile of the surface is usable, and to an ever increasing extent it figures in the support of surface plant and animal life. Of the 55,300 cubic miles of water on the continental surfaces, nearly half is in salty lakes and inland seas (96 percent of this in the Caspian), and slightly more than half is in fresh water lakes, such as the Great Lakes of North America and the rift valley lakes of eastern Africa. About 16,000 cubic miles of water is present within the ground as soil moisture or seepage water, on its way downward to join the ground water, moving laterally down the dip of porous rock layers to emerge as spring water or to supply a swamp, or is being drawn toward the surface by capillary action or the root pressure of plants.

Only about 300 cubic miles of the total world water supply of 326,071,300 cubic miles, hardly 1/1,000,000 of it, is in the rivers and streams coursing through the countryside. This river water in many regions of the world has become so vital to the economic and social systems of man that maintaining and controlling the flow of it, assuring the usable quality of it, and seeing to an equitable allocation of it have become first-order preoccupations of public authorities.

Physical and Chemical Properties of Water

While it is not necessary within the scope of this book to consider in detail the physical and chemical properties of water, it seems desirable to give brief attention to those characteristics which affect its geographic behavior and its economic utility. Physically, water is a liquid under most circumstances, but a fair amount of it is present in a gaseous state as water vapor or in a solid state as snow or ice. Chemically, it is H_2O, an inorganic compound formed by the union of two elements, hydrogen and oxygen.

Until about two hundred years ago, water was considered by scientists to be a simple, uniform substance—that is, one of the elements. In 1766, however, Henry Cavendish, a prominent English scientist, recognized hydrogen as an element and succeeded in burning some he had collected with air so as to form water. A few years later, in 1774, another Englishman, Joseph Priestley, a member of the Presbyterian clergy with scientific interests, succeeded in isolating oxygen in reasonably pure form and in demonstrating with his now famous mint-mouse-candle-glass jar experiment that it can be manufactured by plants under the stimulus of sunlight. The role of sunlight in the process, however, was not clarified by Priestly but was demonstrated seven years later by the Dutch scientist, Jan Ingenhousz.[6]

As a liquid, the form in which it is most widespread, pure water is a transparent and relatively tasteless substance weighing about eight pounds per gallon. It has a solvent action in relation to most solids, a factor which is important in the digestive and growth process of organisms and in the decomposition and redistribution of rock materials over the earth. With the help of humic acid, carbonic acid, and oxygen released by plants, water is able to break down even the hardest of rocks, erode away entire mountain systems, and redistribute the dissolved and fragmented material into vast layers of sedimentary rock. Internal earth forces may then lower these sedimentary layers to greater depths, push them far above the level at which they were formed, fold them, fracture them, and perhaps crumple and thrust them into mountain ranges on which the whole process can begin over again.

The high surface tension of water, greater than that of any other liquid, is utterly essential to the physiological behavior of cells, to the formation of raindrops, and to the form and action of waves. The ability of certain water bugs to walk on the surface of a pond or of a needle to float on the water in a saucer illustrates this characteristic of surface tension. A closely related physical aspect of water is its capillarity, its ability to "defy gravity" by moving in any direction through solids in which there are pores or interstices. It is this quality of capillarity which enables water to disperse itself within the soil and to assist with the movement of life-bearing liquids through the circulatory networks of plants and animals. The phenomenon of osmosis, a characteristic of water as well as of other fluids, by which solutions of unequal strength on either side of a membrane tend to equalize, should also be noted as essential to organic processes.

The transparency of water makes it possible for sunlight to penetrate it for considerable depths, encouraging the growth of plant and animal organisms and warming such a thick mass of it that the oceans moderate the seasonal temperatures of the air above them and even above the windward side of adjacent continents. The high specific heat of water (the amount of heat needed to raise a given volume of it through one degree of temperature) is vital to its moderating influence on temperature change. This high specific heat is responsible for the blanket effect of the clouds and of the water vapor in the atmosphere in preventing overheating of the earth's surface in the daytime and overcooling of it during the nighttime hours.

---

[6]King, *op. cit.*, pp. 15-18.

While water does not change volume with temperature as much as certain other substances, it does change some. Fresh water has its highest density (its least volume in relation to unit weight) at 4 degrees Centigrade, or about 39 degrees Fahrenheit. Below this, especially at the point of freezing, it expands abruptly, causing glass jugs, iron pipes, and even heavy automobile engine blocks to break under pressure which sometimes reaches 30,000 pounds per square inch. Rather curiously, water with the salinity of the oceans contracts in volume with cooling right to the point of freezing, but at that point it expands with abruptness to about 111 percent of its liquid volume.[7] The fact that ice is of lower density than water, and hence will float, affects the transportation relationships of water, the erosive effects of it, the fishing opportunities in it, and a number of other environmental and ecological conditions.

The Hydrologic Cycle

The constant rearrangement of water in space through its evaporation from the ocean and land surfaces; its condensation in the atmosphere to form clouds; its precipitation from the sky in the form of rain, snow, or hail; and its return course to the sea, either by surface streams or by the slow seepage of ground water, is known as the *hydrologic cycle.* While the great bulk of the world's water remains in the ocean, the daily turnover is considerable. An estimated water volume of 280 cubic miles per day is evaporated from the water and land surfaces, more than four-fifths of this from the oceans, and the same amount is returned to the earth as precipitation. Of the 70 cubic miles of water per day which falls on the land, some 50 cubic miles evaporates, and the remaining 20 cubic miles returns to the ocean through runoff or seepage. Subsequent chapters in this book will deal in more detail with the characteristics, behavior, and significance of water in various stages of the hydrologic cycle.

Increasing the Availability and Utility of Water

An important responsibility of man, as his numbers increase and his water requirements mount, is to familiarize himself with the world water supply and to adapt himself and his facilities in such a way that his needs are satisfied. In a small way, he can increase the available supply of water or snow by cloud-seeding to stimulate precipitation.

[7] H. U. Sverdrup, Martin W. Johnson, and Richard H. Fleming, *The Oceans: Their Physics, Chemistry, and General Biology,* Englewood Cliffs, N. J.: Prentice-Hall, Inc., 1942, pp. 68-74.

He can pump water out of the ground, and he can convert water on the ground to more usable form by evaporating, condensing, freezing or thawing it to suit his purposes. Man, of course, cannot make water, but he can move himself, his livestock, his crops, and his factories to it, or (if he is willing to pay the price) he can move the water from the points where nature "delivers it free" to the places where he would rather have it. He can transfer impressive amounts of it through canals and conduits into his reservoirs and specially adapted delivery systems.

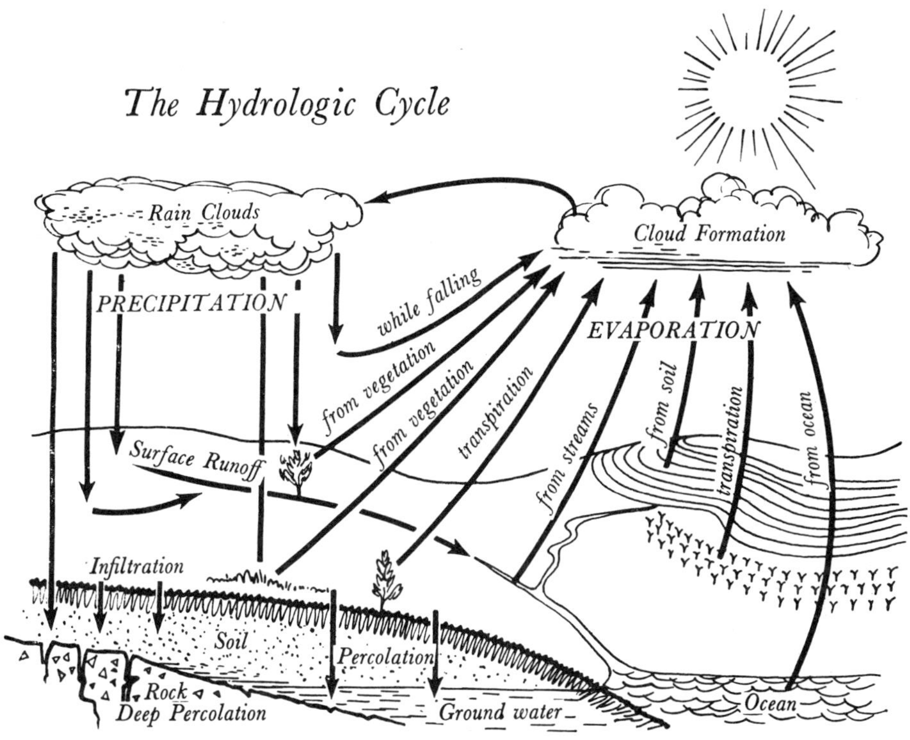

FIGURE 1.1 The Hydrologic Cycle. The time needed for water to complete the cycle from the ocean surface back to the ocean may be as little as a few hours or days or as much as thousands of years. From **Water,** The Yearbook of Agriculture, 1955, p. 42, courtesy U.S. Department of Agriculture.

He can, and often does, pollute it in the process of use, and he can also, when necessary, modify its natural properties or correct his own pollution in such fashion as to improve and maintain its usefulness to him.

Private and Public Agencies Concerned with Water

Water, depending upon the circumstances and the stage in the use cycle, may be either private or public property. It is private property when produced by an irrigation well on a privately-owned farm, and generally it becomes private property when delivered by public authorities to homes, farms, or business establishments. Private individuals and privately-owned business concerns may produce water, transport it, process it, generate power from it, or consume it. They may engage in rain-making, provide insurance against hail or flood damage, produce salt or other minerals from brines, or organize an irrigation district and provide the water for it.

Public agencies at various levels of government are involved in problems of water supply and storage, quality control, distribution, and hydroelectric power production. Some of the earliest known acts of centralized authority, in the valleys of the Nile and the Tigris-Euphrates, were concerned with water allocation. The installation and maintenance of water distribution facilities continue to be a major concern of public authorities, especially at the local level. The resource which once seemed so abundant and so easily available is now expensive to provide and costly to consume. In the writer's home town of Norman, Oklahoma, the sale of metered water is the predominant source of municipal revenue.

In the United States, local governments have most frequently owned and supervised the surface reservoirs, ground-water wells, storage facilities, purification equipment, and distribution lines handling municipal water. Rural water supplies for domestic use, livestock consumption, and small-scale irrigation have commonly been the responsibility of the individual farmer. The federal Soil Conservation Service, however, has constructed thousands of farm ponds on private land and then has turned them over to the landowners. Under state and federal law numerous irrigation districts have been formed, and many water supply systems serving complex clusters of municipalities have been created. Interstate compacts concerning the use of water from boundary rivers or the transfer of water across state lines have been signed by many of the states. City, state, and federal agencies, in order to safeguard public health, have undertaken cooperative pollution control programs, and they frequently work together in the development of water for recreation, consumption, or other beneficial uses.

Navigation on interstate streams is a matter for federal regulation under the constitution. Large sums of money are appropriated by the federal government each year to improve navigation, prevent floods, and assure adequate water supplies for irrigation, industry, and general consumption. The U.S. Geological Survey is responsible for much of the basic research on surface and ground water quantity and quality, for the dissemination of water information through its *Water-Supply Papers* (numbering approximately 2,000), *Hydrologic Atlas* series, and other publications. The U.S. Army Corps of Engineers and the U.S. Bureau of Reclamation have built most of the large dams and reservoirs on the nation's rivers. The U.S. Soil Conservation Service, U.S. Bureau of Public Health, the National Park Service, and several other federal agencies are also participants in the management of the nation's water supply. Water is man's most precious resource, and the provisioning of it for present and future needs requires skilled personnel, a sound philosophy of conservation, adequate financial support, and responsible administration.

Chapter 2

Water in the Oceans and Seas

General Configuration of the Ocean Basins and Bordering Seas

The world's *oceans,* which altogether cover some 71 percent of the global surface, are quite unevenly distributed between the hemispheres. In the northern hemisphere only 61 percent of the surface is occupied by the oceans and *bordering seas,* while south of the equator, water covers about 81 percent of the area. The search by West Europeans for an easy water route to the Orient via a presumed "northwest passage" or "northeast passage" was frustrated for centuries by intervening land and floating ice. In the southern hemisphere at about the same time, the image of a great southern continent (called Terra Australis by imaginative cartographers) was progressively reduced in size. Finally, Captain Cook's voyages of the 1760's and 1770's demonstrated that it was possible to travel around the globe virtually unhindered between 55° and 65° south latitude. Much of the land area of the southern hemisphere is singularly unattractive for human settlement because of ice-caps, mountains, jungles, scrub forests, and barren steppes. Research on the physical characteristics of southern oceans and their marine life has lagged behind that on the more frequently traversed northern oceans, but there is much current interest in understanding the entire global sea, the global atmosphere above it, and the relations between the hydrosphere and the atmosphere.[1]

It has been somewhat traditional in the geographic and general literature to refer to four oceans: Pacific, Atlantic, Indian, and Arctic. Since these so-called oceans are all broadly interconnected, it is perhaps more

[1]Harris B. Stewart, Jr., *The Global Sea,* Princeton, N. J.: D. Van Nostrand Company, Inc., 1963. (Van Nostrand Searchlight Book No. 17) A brief, general introduction to the study of the oceans by the Chief Oceanographer of the United States Coast and Geodetic Survey.

realistic to think of there being just one world ocean, but some of the traditional oceans, by reason of their substantial separation by continental land areas, are more defensible than others. The concept of a fifth ocean, the Antarctic, as a separate water body has been abandoned by professional geographers. Some oceanographers prefer to identify only three major oceans, considering the traditional Arctic Ocean as simply an extension of the Atlantic.[2]

The nomenclature of sub-oceanic water bodies is even more unscientific, and inconsistencies are in evidence almost everywhere. Such terms as *sea, bay, sound,* and *gulf* are used quite indiscriminately in various parts of the globe. The local usages are so firmly fixed in practice that it scarcely seems worthwhile for even such a prestigious office as the U.S. Board on Geographic Names to attempt to redirect the cartographic nomenclature to a more rational system. The Caspian *Sea* and Aral *Sea* are obviously lakes in that they are not directly connected with the ocean at a common level, while *Lake* Ponchartrain and *Lake* Maracaibo, following the same principle, ought to be sounds, bays, or gulfs. In most areas of the world the term *sea* seems to be applied to the larger segments of the water surface which are somewhat separated from the ocean by peninsulas or island groups, but there is no real consistency on this point. The Arabian *Sea* and the *Bay* of Bengal would seem to be similar water features in every important respect, while Hudson *Bay* and the *Gulf* of Mexico are vastly larger than the *Sea* of Marmara or the *Sea* of Azov. Likewise, terms such as *strait, channel,* and *passage* are used interchangeably without consistent connotations of size, shape, or other aspects of configuration.

If the lithosphere (the solid portion of the earth's surface) were as round or spherical as the term suggests, there would, of course, be no dry land. The whole earth would be covered with water to a depth of 2,440 meters, or almost eight thousand feet. While the mean elevation above sea level of the land surfaces is only 840 meters, the mean depth of the oceans and bordering seas is 3,800 meters. The Pacific, which is almost as large as the other oceans combined, has a mean depth of more than 4,000 meters, even including several rather shallow bordering seas. Mean depths of the Indian and Atlantic Oceans are very similar if the adjacent seas are not included. With almost one-fourth of its area in relatively shallow bordering seas, including the Arctic "Ocean," the Atlantic is considered to be the shallowest of the three major ocean basins.

Along some of the continental margins, the descent to true ocean depths is continuous and abrupt, as along much of the west coast of

---

[2]Gerhard Neumann and Willard J. Pierson, Jr., *Principles of Physical Oceanography*, Englewood Cliffs, N. J.: Prentice-Hall, Inc., 1966, p. 15.

TABLE 2.1
Area and Mean Depth of the Oceans and Selected Bordering Seas

Oceans	Area*	Mean Depth
With Bordering Seas		
Pacific Ocean	179.7	4,028
Atlantic Ocean	106.2	3,331
Indian Ocean	74.9	3,897
Without Bordering Seas		
Pacific Ocean	165.2	4,282
Atlantic Ocean	82.2	3,868
Indian Ocean	73.4	3,973
Large Mediterranean Seas		
European-African Mediterranean	2.97	1,458
American Mediterranean (Gulf of Mexico and Caribbean Sea)	4.31	2,174
Arctic Mediterranean	14.06	1,526
Small Mediterranean Seas		
Red Sea	0.44	491
Baltic Sea	0.42	55
Persian Gulf	0.24	25
Marginal Seas		
Bering Sea	2.27	1,437
Gulf of California	0.16	813
Irish Sea	0.10	60

*Area is expressed in millions of square kilometers, and mean depth in meters. One meter is equivalent to 3.28 feet, and one square kilometer to .3861 square miles.

Source: Figures are those of T. Stocks and E. Kossinna, as reported by Neumann and Pierson, *op. cit.*, p. 16.

South America, North America, and Africa. In contrast, the slope of the land surface beneath the waters adjacent to eastern North America, eastern Asia, and northwestern Europe is generally a very gentle one for a considerable distance from shore. The gently-sloping submerged terrace, which ends abruptly in a steep descent to greater depths, is commonly called the *continental shelf.* It is a kind of transition zone between the land and the deep sea. To one examining its profile it seems that there is a little too much water for the *ocean basins* and some of it must "spill over" onto the adjacent coastal plains. The average slope of the continental shelf is about ten feet per mile, while the continental slope, the zone of descent to the deep sea floor, averages approximately 400 feet per mile. Considering the world as a whole, the average distance from the shoreline to the outer margin of the continental shelf is about 40 miles, and the average water depth at the break in slope about 70 fathoms, or 420 feet.[3] For legal purposes the maximum depth of the continental shelf has often been considered to be 100 fathoms, or 600 feet, but such a rounded figure bears little connection with the reality of the break in slope.

The *forces of erosion* (running water, glaciers, wind) which bring about major and minor irregularities on the land surface are generally not operative on the ocean floor. Instead, a slow sedimentation is in process and the bottom of the ocean basins is smoother than the surface of the continents. *Tectonic forces* within the earth, however, cause regional uplifts and subsidences. These create anticlinal and synclinal folds, or they fracture the rocks and bring about differential movement along fault lines. They are not confined in their activity to the present continents and they operate on the sea floor as well. *Vulcanism* (the emergence of molten lava from the rocky depths) is active in the oceans as well as on the continents, and the resultant volcanoes may or may not reach the surface. Some volcanoes, like Mauna Loa in the Hawaiian Islands, grow gradually from frequent non-explosive lava flows, while others, like Mount Pelee in the West Indies (1902) and Krakatoa in the East Indies (1883), may explode with such violence as to destroy life for miles around and modify their whole profile in the process. Even *coral growth* over a long period of time can greatly modify the surface of an island or a reef and may be responsible for navigation hazards to large and small boats alike.

Scientists have been especially intrigued by the submarine canyons, carved somehow across the continental shelves and prolonged into deep water, well beyond the range of possible sub-aereal erosion.[4] Just as finding and exploring the highest mountains on land has attracted the interest of amateur and scientist alike, the search for the deepest spot in the ocean has stimulated many exploratory expeditions. Several dozen *deeps, oceanic trenches,* thousands of feet below the general level of the ocean floor, have been found and explored by means of modern sounding devices. These deeps are ordinarily adjacent to and parallel with zones of crustal upfolding, often reflected in an arc of mountainous islands or a chain of coastal sierras like the Andes of South America.

For many years it was assumed that the deepest of the ocean deeps was the Philippine or Mindanao Trench off the east coast of Mindanao, with water depths of just under 35,000 feet. The vigorous oceanographic research conducted by a number of nations since World War II has resulted in several record probings. In late 1952 the *Horizon,* a research vessel of the Scripps Institution of Oceanography, encountered

---

[3]Stewart, *op. cit.*, pp. 32-33.

[4]Francis P. Shepard, *Submarine Geology,* N. Y.: Harper and Row, 2d ed., 1963.

a depth equal to that near Mindanao in the Tonga-Kermadec Trench northeast of New Zealand. In 1960 the United States Navy's bathyscaphe, *Trieste,* with two men aboard reached a depth of 35,800 feet in the bottom of the Mariannas Trench, not far from the spot where a year earlier the Soviet Russian ship *Vityaz* reported in the same trench a depth of 36,204 feet southwest of the island of Guam.[5] The British survey ship H.M.S. *Cook* reported in 1962 that it had found a depth of 37,782 feet in the Mindanao Trench, but this report later proved to be in error. There is no particular reason to believe that still greater depths may not be found in the future. While the ocean trenches are being prodded to even greater depths, other survey expeditions are recording the details of enormous *submarine ridges.* The long-known Mid-Atlantic Ridge and the Lomonosov Ridge in the Arctic basin have been recently charted by Soviet and American scientists with the aid of nuclear-powered submarines.

The Science of Oceanography

Adventurous men have been going "down to the sea in ships" since prehistoric time. The earliest inhabitants of North and South America seem to have arrived by crossing the narrow but stormy Bering Strait, although some investigators would have them crossing the Pacific nearer the equator, island-hopping in Kon-Tiki-like rafts until they saw the towering Andes and decided to come ashore. Phoenician and Greek sailors hundreds of years before the time of Christ were trading, raiding, and venturing throughout the Mediterranean Sea. They pushed beyond the fabled Pillars of Hercules, the modern Strait of Gibraltar, to explore the coasts of Africa and northwestern Europe. Arab traders were well acquainted with the Asian and African shores of the Arabian Sea before Vasco da Gama finally made it around the Cape of Good Hope and reached India. School children everywhere are familiar with the exploits of the Vikings and of men like Christopher Columbus, Amerigo Vespucci, Ferdinand Magellan, and Henry Hudson, who sought new routes and found new lands previously unknown to the Europeans.

Advancement of navigation and the progress of discovery were facilitated by the development of technology and instrumentation, some comprehension of the behavior of the stars, the moon, and the earth in space; the invention of the gnomon, the astrolabe, the back-staff and the sextant; the use of the rectangular grid; the magnetic compass, the chronometer, and the ephemeris. For centuries, however, it was necessary to rediscover faraway islands and harbors over and over as

[5]Stewart, *op. cit.,* pp. 22-25.

man developed what E. G. R. Taylor has called "the haven-finding art." But by the second half of the eighteenth century, expeditions like those led by La Perouse and Captain Cook could chart new lands and expect them to remain charted.[6] Governments of seafaring naitons by the eighteenth and nineteenth centuries were establishing official hydrographic offices to study the characteristics of the oceans and seas and to chart in detail the coastlines and harbor approaches. France established such an office in 1720, Denmark in 1784, Great Britain in 1794, the United States in 1830, Japan in 1871, and Germany in 1875.[7] The Russian government was a little slow in promoting oceanographic research, but the Soviet Union is now widely regarded as a leader in this field of scientific activity.

The modern field of oceanography devotes major attention to the physical and chemical characteristics of the oceans, the behavior of ocean water, and the organic forms which occupy it. It is an aspect of geography, at least in the broader usage of that term, and is obviously closely related to and overlapping with the sciences of meteorology, geology, zoology, and botany. Sir William A. Herdman has described and evauated in some detail the work of several of the leaders in nineteenth- and early twentieth-century oceanography,[8] and many of the textbooks devoted to general oceanography include a chapter or two on the history of the field.

Physical oceanography owes much of its initial impetus and pattern of organization to Matthew Fontaine Maury (1806-1873), an American naval officer whose observations and compilations on winds, tides, and currents were highly praised, and whose book, *The Physical Geography of the Sea,* became, in effect, the first textbook on oceanography.[9] Maury was largely responsible for organizing an international conference on the scientific study of the oceans in Brussels in 1853, at which representatives of ten maritime nations, later joined by nine others, agreed to promote a uniform system of observations at sea. A contemporary of Maury, Edward Forbes (1815-1854), a self-trained

---

[6]E. G. R. Taylor, *The Haven-Finding Art: A History of Navigation from Odysseus to Captain Cook*, London: Hollis and Carter, 1956.

[7]Gunter Dietrich, *Ozeanographie*: Physische Geographie des Weltmeeres, Braunschweig: Georg Westermann Verlag, 2d ed., 1964, p. 6. It should be noted here that Dietrich's 1830 date for the United States refers to the establishment of the Depot of Charts and Instruments in the Department of the Navy. The work of the Depot was absorbed by the Hydrographic Office of the Navy, created in 1866. The United States Coast (and Geodetic) Survey, whose major functions include the preparation of detailed maps of the coast and coastal approaches, dates from 1807. The Coast and Geodetic Survey was recently merged with the United States Weather Bureau in the Environmental Services Administration.

[8]Sir William A. Herdman, *Founders of Oceanography and Their Work: An Introduction to the Science of the Sea,* London: Edward Arnold, 1923.

[9]Matthew F. Maury, *The Physical Geography of the Sea,* New York: Harper, 1855.

natural scientist in the tradition of Alexander von Humboldt, Charles Lyell, Alfred Russell Wallace, and Charles Darwin, was a major pioneer in biological oceanography. Born on the Isle of Man and later a distinguished Professor of Natural History at the University of Edinburgh, Forbes made clever use of the dredge in sampling and describing life zones at various depths in the ocean. He eventually published a map of the distribution of marine life throughout the world. Other prominent pioneers in nineteenth-century oceanography include Wyville Thomson and John Murray, associated with the 69,000 mile expedition of the specially-equipped British research vessel, *H. M. S. Challenger* (1872-1876). The fifty large volumes of *Challenger Reports,* issued over a fifteen-year period following the expedition, present a wealth of detail on ocean floor contours, sea bottom sediments, water temperatures, currents, and marine life.

Oceanographic and hydrographic research has been furthered in the twentieth century by the work of government agencies in many countries and by the efforts of institutes and special research centers supported by universities, business corporations, private foundations, and wealthy individuals. Major expeditions have been mounted by the Soviet Union, the United Kingdom, the United States, France, Germany, Belgium, the Netherlands, the Scandinavian countries, and Japan, and much new data has been assembled. Several international organizations are active in the field of oceanography. An International Council for the Exploration of the Sea was organized in 1902, and an International Ice Patrol was established in 1914. The International Hydrographic Bureau, formed in 1921 as an outgrowth of the First International Hydrographic Conference held in London two years earlier, has a membership of 34 nations. It exists for the purpose of facilitating the exchange of all types of information in the field of hydrography. Major conferences of the membership are held every five years. The International Association of Physical Oceanography, with its headquarters in Göteborg, Sweden, originated in 1919 as a subdivision of the International Union of Geodesy and Geophysics and has a major convention each three years in association with its parent organization. Since the successful international effort in scientific cooperation known as the International Geophysical Year (1957-58), which included substantial attention to oceanographic studies, the United States and the Soviet Union have each played host to a major International Oceanographic Congress. The first was held in New York City in 1959 and the second in Moscow in May and June, 1966. The Moscow congress, organized by the U.S.S.R. Academy of Science had the support of UNESCO, the UN Food and Agricultural Organization, the World Meteorological Organization, the International Atomic Energy Commission, and the Scientific Committee on Oceanic Research.

In the United States, the Scripps Institution of Oceanography at La Jolla, California (founded 1922), and the Woods Hole Oceanographic Institution at Woods Hole, Massachusetts (1930), are outstanding centers of oceanographic research, active in training scientists and conducting studies in physical and biological characteristics of the sea. Fine work is also being done in the field at a number of universities, including the University of Washington, the University of Miami, and Texas A. and M. University.

The accumulation of knowledge concerning the characteristics of the oceans is being pursued in a great variety of ways. It is reported that in 1964 there were about 340 ships in the oceanographic fleet of the world, varying in length from less than 50 feet to more than 400 feet. One of the newest of these vessels is the *Atlantis II* operated by the Woods Hole Oceanographic Institution. It is 210 feet long, with a beam of 44 feet and a draught of 17 feet, a cruising speed of 12 knots, and a range of 8,000 miles. The *Atlantis II* has precision echo-sounding equipment and a variety of other technical gear, underwater lights, and an underwater observation chamber with six viewing ports, a complete system of air-conditioning, and comfortable quarters for 25 scientists and a crew of 28.

In addition to the specially-equipped oceanographic vessels, some use in oceanographic research is made of orbiting satellites, aircraft, hover craft, hydrofoils, fixed towers and platforms, buoys and floats, submarines, diving research vehicles, bathyscaphs (depth ships), and even buoyant underseas laboratories in which men work for days at a time at depths of up to several hundred feet. Although the idea still seems a bit extreme, popular science magazines are beginning to carry articles about the prospect of permanent habitation under water—of real "cities under the sea." Since the days when Magellan and his crew measured ocean depths with a piece of metal and a plumb line, and Benjamin Franklin set corked bottles afloat in the Atlantic Ocean to study the currents, the instrumentation used in oceangraphic research has grown increasingly complex, accurate, and expensive. The Museum of Oceanography in Monaco, which the writer has visited on two occasions, has a fascinating historical collection of oceanographic instruments.

Depth, Salinity, and Temperature Characteristics

Depth surveys of the world's oceans, beginning with those of the *Challenger* expedition, have made it possible to portray in considerable detail the relief of the ocean floor. A spectacular interpretive diagram of the ocean floor, making use of some of the most recent IGY research, was published in color in the issue of *Life* magazine for November 7, 1960.[10] This diagram shows large areas of relatively even profile interrupted by pronounced swells, ridges, clefts, domes, and peaks.

From the *Life* map it is obvious that broad sectors of all three major oceans have depths of from two to four miles in the areas known as *abyssal plains*. Rising above the abyssal plains are extensive submerged ridges, often of remarkable linear extent and with surprisingly steep marginal slopes, while below the plains, most strikingly in the western Pacific area, are a number of arcuate deeps which extend to depths of five to seven miles. The ocean floor is spotted with mountains which when submerged, as most of them are, are called *seamounts*, but which in places appear above the surface as large islands like Iceland or the Hawaiian group, or as small ones like Palmyra, Pitcairn, or Tristan da Cunha.

Almost everyone knows that ocean water is salty and unfit to drink. It cannot be used for direct consumption by either people or domesticated animals without *desalinization*, still a rather expensive process. It can seriously damage soils on a low coastal plain if it is allowed to flood in upon or seep into areas planted to farm crops. The term *salinity* as usually applied to ocean water is somewhat misleading, as it includes many dissolved substances in addition to the sodium chloride ($NaCl$) which is common table salt. Since the evaporation from the surface involves only the removal of "distilled" water, the return flow of streams and ground water is constantly adding mineral matter from the rocks with which this water has been in contact.

The present salinity of the oceans is about 3.5 percent by weight, or 35 parts per thousand, which is the usual way of expressing it. The ten most common elements found in ordinary sea water are:[11]

Elements	*Parts per Thousand*
Chlorine (Cl)	18.98
Sodium (Na)	10.56
Magnesium (Mg)	1.27
Sulphur (S)	0.88
Calcium (Ca)	0.40
Potassium (K)	0.38
Bromine (Br)	0.065
Carbon (C)	0.028
Strontium (Sr)	0.013
Boron (B)	0.005

[10]"The Structure of a Not-So-Solid Earth," Part I of a series entitled "The New Portrait of Our Planet," *Life*, vol. 49, November 7, 1960, pp. 74-101.

[11]Warren E. Yasso, *Oceanography: A Study of Inner Space*, New York: Holt, Rinehart, and Winston, Inc., 1965, p. 22.

These ten elements comprise about 33.5 parts per thousand of the total dissolved salts, and the remaining 1.5 parts include many other elements, among them copper, aluminum, manganese, uranium, gold, and silver. It has been calculated that a cubic mile of sea water contains 23 pounds of dissolved gold, worth almost $13,000 if it could be recovered cheaply enough to make it worthwhile. If all of the salt in the ocean were precipitated and spread uniformly over the ocean basin area, it would have a depth of almost 200 feet. There is no shortage of salt in the world!

While the mean salinity of the three major oceans varies only slightly from the 34.72 parts per thousand calculated as the mean of all ocean water, the actual salinity at a specific spot might vary from about 33.0 to 37.0. Surface salinity in the open ocean tends to be slightly higher in areas of warmer air temperature and much sunshine, while near the continental margins, especially near the mouths of major rivers, it can be much lower. The Mediterranean Sea has a mean salinity of 38.5, and the Red Sea, with its high surface evaporation and negligible increment of fresh water, normally has salinities above 40.0. The nearly enclosed Baltic Sea, on the other hand, with a frequent cloud cover and a large number of rivers entering it, has salinities between 20.0 and 25.0. The Gulf of Bothnia arm of the Baltic has a mean salinity that is regularly below 10.0.[12]

At any given latitude, whether the earth's surface is land or water, the amount of solar insolation reaching the surface is approximately the same. Because of the transparency, mobility, and high specific heat of water, however, the temperature characteristics of the oceans are very different from those of the land and the air over it. Over the open sea, air temperatures are never greatly different from those of the water surface. Temperatures of the water surface, cooled somewhat in the process of evaporation, seldom change more than 10° F. with the seasons, even in middle latitudes, or more than 1° F. between day and night. Beyond a depth of about 200 meters there is no temperature change of a seasonal nature. The temperature conditions just described would not necessarily prevail in shallow water near the margin of the land nor in those areas where the change of seasons brings a change in ocean currents or vertical movements of the water.

The average temperature of the entire ocean surface has been calculated to be about 63° F. It is somewhat higher than that in the northern hemisphere (67° F.) and somewhat lower in the southern (61° F.). The Pacific, with its greatest breadth near the equator, has the highest average surface temperature of the three great oceans. Near

[12]Hugh J. McLellan, *Elements of Physical Oceanography,* Oxford: Pergamon Press, Inc., 1965, p. 17.

the equator the surface temperature is always about 80° F. or slightly higher, and in Arctic and Antarctic waters it is about 29° F., or 3° below the freezing point of fresh water. Deep water is always cold, with bottom temperatures generally between 39° F. and 29° F.[13] Unlike fresh water, sea water becomes heavier as it is cooled to the point of freezing and tends to sink to lower levels. Water is slightly compressible, and at great depths freezing is retarded by the heat of compression. Because expansion occurs with freezing, sea ice floats readily and thickens downward from the surface when the air temperature above it is sufficiently cold. Icebergs born in either the Arctic or Antarctic may be carried hundreds of miles by ocean currents, well into middle latitudes. The sinking of the *Titanic* on its maiden voyage in the North Atlantic on April 15, 1912, as a result of full-speed collision with an iceberg caused a loss of more than 1,500 lives. The *Titanic* disaster had a strong influence in the formation of the International Ice Patrol in 1914.

Circulation and Movement of Ocean Water

Some years ago, Dr. F. Kenneth Hare, a well-known British geographer, wrote a book entitled *The Restless Atmosphere.* The adjective "restless" might be applied with equal propriety to the world's oceans. Sea water is in constant motion, often even several different kinds of motion at the same time. The surface is commonly involved in the oscillatory movement of waves and tides, motions which are particularly apparent along the shore, but at the same time it is involved in "movements of translation" which displace large volumes of water either vertically or horizontally. The vertical displacements are the result of disequilibria consequent upon differences in salinity or temperature, both of which affect the density. Colder water tends to settle beneath warmer water, and saline water tends to remain under fresh water, as where a river enters the sea, although in both types of situations considerable mixing occurs. Massive horizontal transfers of water, known as *currents,* occur in the oceans because of the influence of wind, the rotation of the earth, the configuration of the continental shorelines, and the continuing effort of counter flows to maintain a "level" surface on the curving sphere.

The general pattern of surface water circulation is not excessively complicated. The trade winds, which blow with some persistence from the northeast in the northern hemisphere and from the southeast in the southern hemisphere, drive water generally westward in the Pacific and Atlantic, between 30° north and 30° south. In partial com-

[13]R. E. Coker, *The Great and Wide Sea,* Chapel Hill: University of North Carolina Press, 1947, pp. 94-99.

pensation there develops a west to east counter-current in a narrow belt approximately along the equator. When the main west-moving current in the Atlantic strikes the east coast of South America the configuration of the coast diverts most of the water northwestward toward Florida where the term Gulf Stream is applied to it and then northeastward into the north Atlantic where it spreads out and slows down as the North Atlantic Drift. Some of this water warms the shores of Scotland, Norway, and Novaya Zemlya, and then gets involved in the rotary circulation of the Arctic sea. Another major segment of it turns south along the western coast of Portugal, Morocco, and Mauritania to rejoin the western drift of subtropical water driven by the trade winds. This clockwise rotary motion of surface water in the northern basin of the Atlantic is matched with a similar clockwise circulation in the Pacific Ocean north of the equator. In the southern Atlantic basin and in the Pacific basin south of the equator there is a rotary circulation in the opposite (*i.e.*, counter-clockwise) direction. The persistence of westerly winds throughout the year between about 45° and 65° south latitude and the lack of any sizable land obstruction cause a west-to-east water circulation in the southern seas, or what has sometimes been called the Antarctic Ocean.

The powerful seasonal reversal of wind direction over the Indian Ocean, known as the summer and winter monsoon, is caused by the winter high and the summer low pressure development over Asia. This reversal of the winds also affects the circulation of water and brings about a less persistent pattern of currents than is characteristic of the Atlantic and the Pacific.

Somewhat without regard to its specific temperature, water that is moving toward the equator is designated as a *cold current,* while water moving away from the equator is called a *warm current.* The most prominent cold currents would include the northward-flowing Peruvian (or Humboldt) Current along the west coast of South America, and the Benguela Current along the coast of southwestern Africa, the southward-flowing Labrador Current off eastern Canada and New England, the Oya Shio Current along the coast of eastern Siberia, the California Current along the coast of California and the peninsula of Lower California, and the Canaries Current along the coast of northwestern Africa. The major warm currents are the Brazil Current, the Mozambique (or Agulhas) Current, and the East Australian Current in the southern hemisphere, and the Kuro Shio-North Pacific Current and Gulf Stream-North Atlantic Drift in the northern hemisphere. The general pattern of surface circulation in the Arctic Ocean is a clockwise rotation, as verified by Fridtjof Nansen in the drift of the *Fram* (1893-96) and by many later observers.

The rate at which water moves in most ocean currents is quite slow, commonly less than one knot (one nautical mile or 6,076 feet per hour), although where it is constricted by pressure against a land mass or between two land masses the velocity may be greatly increased. The Gulf Stream off Florida is moving at three to four knots and the transport of water here has been estimated at about 26 million cubic meters per second, or something like one thousand times the volume of the Mississippi River at normal stage.[14] It is perhaps not surprising that studies are under way, with huge funnel-like experimental turbines, to determine the feasibility of some day developing electric power from the Gulf Stream. Another type of hydroelectric power experiment has used the contrast in temperature between warm surface water and the cold water a few hundred feet beneath the surface in the generation of steam power.

Water movements that are primarly vertical are described in some detail in the literature of oceanography. *Subsidence* of surface water takes place under a variety of conditions, generally as a result of contrasts in temperature or salinity but also where, due to winds or tidal influences, a mass of surface water moves against a coast. Where winds are persistently offshore, driving surface water away from the land, there is *upwelling* of colder water from depths of up to 200 or 300 meters.[15] Enough water is involved in this upwelling to create abnormally low surface temperatures for the latitude and even to modify the local climate. The Atacama Desert coast of western South America and the almost equally dry northwestern coast of Africa in Morocco and the Spanish Sahara, are two well-known areas where cold water from lower levels rises to the surface.

A special type of subsurface current known as a *turbidity current* has been postulated by R. A. Daly, P. H. Kuenen, F. P. Shepard, and others as an aid to explaining the presence of coarse sediments far out from shore and perhaps as a major explanation for many of the *submarine canyons* which cut deeply into the continental slope. Such currents, presumably set in motion by submarine landslides or earthquakes, start a downslope movement of water with a heavy suspended load, and the added density resulting from the suspended load forces an increase in velocity which, it is thought, can reach 25 or even 50 knots. Turbidity currents with such velocity would have tremendous erosive power, and investigators have concluded that they were probably responsible for the succession of cable breaks in a seaward direction

---

[14] *Ibid.*, p. 127.

[15] Cuchlaine A. M. King, *An Introduction to Oceanography*, New York: McGraw-Hill, Inc., 1963, p. 117.

which occurred after the earthquake of 1929 on the Grand Banks off Newfoundland.[16]

The periodic rise and fall of water along the seashore known as the *tide* has fascinated simple observers and learned scholars since ancient times. Basically, the tide is the result of the gravitational attraction of the moon and, to a lesser degree (because it is 400 times as far away) of the sun on the surface of the earth. Even the land surface responds some to this tidal pull, as much as several inches in the deep interior of the continents, according to IGY measurements. But it is along the ocean margin that it is easiest to observe. The force of gravity, which has been understood with some degree of precision since the time of Sir Isaac Newton (1642-1727), attracts the mobile water of the oceans in such a fashion as to cause a slight bulge on the side nearest the moon and a second bulge on the exact opposite side. As the earth rotates and the moon itself revolves around it, the position of the two bulges changes, setting in motion a kind of cyclical rise and fall of the water at any given place. Since points on the earth's surface would be affected by one of the tidal bulges approximately each twelve hours, there is a tendency for two high tides and two low tides to occur each day. The fact that the lunar day is about fifty minutes longer than the solar day is responsible for making each of the two daily high tides occur some twenty minutes later than the corresponding high tide of the previous day. The so-called *spring tides,* relatively high range ones, occur when the gravitational attraction of the moon and sun is along the same axis. The term *neap tides* is applied to those lower range ones which occur when the lunar and solar attraction is along two axes at right angles to each other. The time interval between two spring tides or two neap tides is about 29 days, with the changes following the phases of the moon. Spring tides are associated with the times of new moon and full moon, while neap tides occur with the intermediate phases.

Changes in the relative position of the plane of the ecliptic of the earth and the moon, the rotation of the earth and the effect of the Coriolis force on the direction of surface currents, the distance of a specific point on the shore from the amphidromic point in the ocean around which the tidal effect occurs, the trend or direction of the shoreline, and the indentations of the shoreline—are some of the reasons that the tidal range is very different from place to place. The actual interval between successive high and low tides at different locations is far from uniform. Enclosed or semi-enclosed seas or bays, such as the Mediterranean Sea or Hudson Bay, usually have a lower tidal range than the shore of the open ocean, perhaps only a foot

[16]Shepard, *op. cit.,* pp. 138-140 and 335-347.

or two. On the other hand, funnel-shaped entrants to the land from the open ocean, such as the Severn River estuary or the Bay of Fundy, may have a daily tidal range of 25 to 50 feet. A hydroelectric power station making use of the tremendous energy of the incoming and outgoing tide has been installed on the north coast of France near St. Malo.

Waves and *swells* in the ocean are basically due to wind action, although not necessarily to wind action at the time and place of observation. Wave height (vertical distance from trough to crest) is usually only a few feet and seldom exceeds forty feet; although the water from a wave striking the bow of a ship or breaking on a rocky shore may splash as high as a hundred feet. Although the wave motion, or wave profile, moves forward at a velocity reflected in the wave length, a particular segment of the water moves forward hardly at all, except along the shore where the rotating mass drags on the bottom and the wave crest rushes up the sloping beach. While a surfboat can move forward on the wave motion itself, a floating log in a stormy sea will do little but bob up and down in place as the waves pass under it.

The presence of ice in the sea, in polar, subpolar, and even middle latitudes, affects its trafficability and the approachability of adjacent coasts. It is at the same time a cause and an effect of high latitude climates. The isolated iceberg which floats into middle latitudes may catch a ship unaware and cause extensive loss of life and property, as in the case of the hapless *Titanic.* The cold temperature of high latitude waters has a pronounced effect on the marine organic life and on a variety of human interests including salt-water bathing.

While it is impossible to measure precisely, it has been estimated that drifting ice covers more than six percent of the entire ocean surface.[17] This ice is of two general types, *icebergs* and *sea ice* or *pack ice.* The icebergs are fragments of *glaciers* formed on the land which move slowly seaward by gravity and are forced to *calve,* or break loose in segments, by the buoyance and movement of the sea water into which they are projected. Icebergs vary greatly in size at the point of their formation, occasionally originating as blocks 20 to 40 miles in length. More often they are less than a mile long and stand not more than 30 or 40 feet above the surface. Unless they are trapped in a largely frozen basin, such as the Arctic Sea, they are carried equatorward by cold currents and are melted within two years. Almost all of the icebergs of the southern hemisphere are formed at the margin of the Antarctic continent or its shelf ice extension, while in the northern hemisphere there are many source areas, with Greenland, Franz Josef Land, Novaya Zemlya, and Alaska having particular importance. On a trip through the Strait of Belle Isle, north of Newfoundland, during

[17]King, *op. cit.,* pp. 110-113.

FIGURE 2.1 Pack Ice and Icebergs. Courtesy U.S. Coast Guard. The 269-foot icebreaker "Westwind," shown here in the frozen waters off Thule, Greenland, was assisting a Danish freighter trying to make its way with supplies to a remote Eskimo village. Picture taken on September 25, 1964.

the second week of June, 1963, the writer saw many small icebergs which had been carried down through Baffin Bay and the Davis Strait by the Labrador Current. Only a small fraction of the icebergs formed in high latitudes reach as far equatorward as 45° N. or S., but occasionally they survive to within 35° of the equator.

Sea ice, or pack ice, is the floating ice which forms directly on the surface of the ocean and bordering seas in high latitudes. It begins to form as a kind of slush when crystals of frozen fresh water separate themselves from an increasingly concentrated salt solution. If the cold temperature persists long enough, the whole mass solidifies and thickens on the under side. The thickness of sea ice seldom exceeds 30 feet, although the winds or currents bringing masses of it together may occasionally build up *pressure ridges* to a thickness of 60 to 100 feet. Old sea ice, which survives from season to season, covers about 70 percent of the entire Arctic basin, and much new ice forms during the winter season. A considerable amount of sea ice survives the summer season among the islands of the Canadian archipelago, its exact position, as in the more open Arctic basin, depending upon the strength and persistence of prevailing winds. Sea ice extends far offshore from the Antarctic continent, especially in the two major indentations of the continent, the Ross Sea and the Weddell Sea.

Storms at sea have been a favorite subject of several novelists, and they provide vivid memories for countless sailors and ocean-going tourists. The howl of the wind in the rigging of a Clipper ship making its way around Cape Horn; the determined dip and heave of a giant modern liner in the face of waves which splash over the deck and drive the passengers to the bar indoors or to their bunks below; the lonesome terror in the fisherman's dory which, caught in a storm, loses contact with the mother ship; the sickening crunch of metal and the dangerous oil slick which spreads outward from the tanker driven onto the rocks by the might of the wind—these are the stuff of fiction for the landlubber and the reality of existence for those who spend their lives at sea. Storms are a reaction to prolonged high winds or abrupt pressure changes over the water, and they may occur in almost any part of the ocean. They are, however, probably least frequent in equatorial waters and most frequent in the waters of the North Atlantic basin, the northern part of the Pacific basin, and the belt of continuous westerly winds which encircles the globe in the southern hemisphere. The *tropical hurricanes* of the western Atlantic basin and the *typhoons* of the Pacific basin are destructive storms of great violence which involve the rapid rotation of air around centers of very low air pressure.

Marine Plant and Animal Life

Life on earth seems to have originated in the ocean, and there is at present a remarkable diversity of both plant and animal life in it. The fact that the total volume of organic life in the world's oceans is much greater than the volume on land is explained in that a much greater vertical zone of the water is habitable. The amount of organic life which can be supported in the water is dependent upon a variety of factors, including light, temperature, salinity, oxygen, carbon dioxide, phosphates, various nitrogenous compounds, and certain toxic materials such as hydrogen sulfide. The sediment at the bottom of the ocean, in addition to materials brought in by rivers, includes great depths of ooze formed from the incompletely decayed remains of diatoms and other microscopic marine life. The term *plankton* is applied to the microscopic organic life in the oceans, whether of plant (*phytoplankton*) or animal (*zooplankton*) origin.

Plant life, being dependent upon sunlight for photosynthesis, is largely limited to the upper 250 feet or so of the ocean. There are hundreds of varieties and forms of microscopic algae, many of them with siliceous coatings which resist decay as the dead remains sink to the bottom and form ooze. The volume of phytoplankton present, which has much to do with the ability of the water to support animal life, is greatest in (a) areas of upwelling water, as off the subtropical west coasts of continents; (b) high latitude waters, especially during the summer season when there is an abundance of sunlight; (c) coastal regions, particularly where nutrients are brought in by rivers from off the land and where waves and currents stir the water and so help to aerate it; and (d) areas where there is a convergence of warm and cold water, as where ocean currents from opposite directions approach each other. Because of the important role of phytoplankton in feeding small animal organisms, upon which larger fish feed, the major commercial fishing banks of the world tend to be in locations such as those just described.

While most of the plant forms which grow in the ocean are of microscopic or near-microscopic size, a number of them are much larger. Some types of brown algae, such as rockweed and kelp, are very coarse and form thick mats over large areas of shallow water near the shore. The sargasso weed, *Sargassum,* which has little air bladders enabling the plant to float on the surface of the water, has accumulated to such an extent in the southwestern part of the North Atlantic basin that the term "Sargasso Sea" is applied to a segment of the ocean and is shown prominently on many atlas maps. Eel grass, *Zostara,* is another large sea plant which grows in shallow water and which is an important food supply for fish. Several well-known land plants grow along shorelines in water of ocean-like salinity, among them the mangrove and the coconut palm.

As in the case of plant life, the animal life in the ocean includes forms varying in size from microscopic to huge; from single-celled protozoa to giant whales weighing up to a hundred tons. In general, the same physical conditions which encourage a vigorous plant life in the sea also support an abundant animal life. The small animal forms, such as the single-celled *Radiolaria* and *Foraminifera* and the visible but still tiny *Copepoda,* constitute the zooplankton whose degree of abundance has much to do with the density of larger forms. Some sea animals live constantly within a few feet or yards of the surface, while others habitually live at much greater depths far below the zone penetrated by sunlight to which marine plants are confined. Some are permanently attached to the sea floor, such as the crinoids, sea anemones, and adult corals, while other bottom dwellers, like the starfish and mollusks, are able to move slowly about in search of their food supply.

Special Characteristics of the Major Ocean Basins

As was indicated earlier in the chapter, it is becoming increasingly common among oceanographers to refer to only three major oceans, the Pacific, the Atlantic, and the Indian, to regard the Arctic "Ocean" as a sea, and to ignore the Antarctic "Ocean" as not having sufficient physical separation to be considered a separate water body at all. The divisions between even the major recognized oceans are not precise since they have broad, open connections with each other.

The Pacific Ocean with its bordering seas covers almost one-third of the earth's surface. It is nearly surrounded by a rim of young, rugged mountains, many of which are still subject to vulcanism or frequent earthquakes. While it is larger in area than the other oceans combined, it is only slightly deeper. Except for the Aleutian and Hawaiian chains, the northeastern half of it (approximately northeast of a line from Tokyo to Valparaiso, Chile) is remarkably free of islands. The southwestern half is studded with islands, including large ones like New Zealand, New Guinea, Mindanao, Luzon, and Honshu, and thousands of small ones, some of them well-known like Tahiti, Wake, Saipan, and Iwo Jima. Despite its name, which seems to derive from impressions of the Magellan expedition that was delayed for weeks by the equatorial doldrums, it is not significantly less stormy than the others. The tropical hurricanes of the western Atlantic are matched in intensity and ferocity by the typhoons of the western Pacific.

The Atlantic Ocean is about half the size of the Pacific. In contrast to the Pacific, which has its greatest width approximately along the equator, the Atlantic narrows to a kind of waist in the vicinity of the equator, where South America and Africa approach to within 1,600 miles of each other. For some purposes, because of this narrowing, it is convenient to recognize a North Atlantic basin and a South Atlantic basin, although deep water connects the two. Particularly striking on a relief map of the Atlantic floor is the Mid-Atlantic Ridge which extends as a submerged plateau from the Arctic to the Antarctic for a distance of about 10,000 miles. Its dominant trend, paralleling the continental margins, tends to support the theory of Alfred Wegener and others that the western hemisphere continents at one stage of geologic history broke away from the Eurasia-Africa land mass and drifted apart to their present position. The general configuration of the continental shoreline on either side of the Atlantic strongly supports the "continental drift" theory, although examined in detail the theory presents a number of unsolved geomorphic problems. Most of the larger islands of the Atlantic are along its periphery, although several small ones (Azores, St. Paul's Rocks, Ascension, Tristan da Cunha) are formed along the Mid-Atlantic Ridge.

The Indian Ocean, which lies mainly in the southern hemisphere, is set apart from the rest of the "world ocean" by the land masses of Africa, Asia, Australia, and the Sunda Islands. It has a broad connection with the other major oceans between the Antarctic Circle, which marks the approximate northern extent of the Antarctic continent, and about 35° south latitude at the southern extremity of Africa and Australia. Perhaps because of its remoteness from the countries which lead in oceanographic research, but also because it has been less fought over and fished in than the Pacific and the Atlantic, it is less well-known than the other two. On close examination, however (and, incidentally there has been a great deal of interest in Indian Ocean research since World War II), this ocean presents complexities enough to challenge a prolonged research effort. As in the case of the Pacific and Atlantic, the larger islands of the Indian Ocean are on or near the periphery (Madagascar, Ceylon, Sumatra, and Java) while only smaller islands and island groups (Seychelles, Maldives, Chagos, Mauritius, Reunion, Cocos, and others) are in the mid-ocean. One of the curiosities of the Indian Ocean is the seasonal change in direction of currents in the northern part of the basin in conformance with the seasonal reversal of the monsoon winds.

The Arctic Sea, as a polar projection map portrays most visibly, opens to the Pacific only through the fifty-mile wide Bering Strait, but it has a broad opening to the Atlantic and, in a sense, can be regarded as an extension of the Atlantic. The Atlantic and Arctic basins are substantially separated at deeper levels by the Wyville Thomson Ridge, which extends northwestward from Scotland to the Faeroes, and by the *rises* on the ocean floor to the east and west of Iceland. There is a considerable exchange of surface water between the North Atlantic and Arctic basins, however, by means of the North Atlantic Drift, which carries relatively warm water to the northern extremity of Norway and even to western Spitsbergen, and by the East Greenland and Labrador currents which carry icy water south to Newfoundland, Nova Scotia, and northern New England. Efforts to find a usable route to the Orient by either a "Northwest Passage" or a "Northeast Passage" across the Arctic were frustrated by sea ice until the late nineteenth century, and even now such routes are hardly practicable from the commercial standpoint. The recent discovery of oil at Prudhoe Bay along the north coast of Alaska, however, and the successful voyage of the reinforced tanker, the SS *Manhattan,* through the ice-bound Northwest Passage in the autumn of 1969, arouse new hope that a way can be found to make this route feasible. Important knowledge about the circulation of water and sea ice in the Arctic basin was gained by the expeditions of Fridtjof Nansen in 1893-95 and Ivan Papanin in 1937-38. Since World War II a vast amount of Arctic basin research has been conducted through aerial surveys by United States, Canadian, and Russian flyers, by the American *Nautilus* expedition (1957), and by other submarine research under the Arctic ice. The greatest depth so far encountered in the Arctic is only about half that of the Mariannas Trench in the Pacific.

Special Characteristics of Selected Bordering Seas, Bays, and Gulfs

While at least gross similarities exist between the major ocean basins, the bordering seas, bays, and gulfs present striking contrasts in their shoreline and bottom configuration and in such characteristics as temperature, salinity, circulation, tidal phenomena, and organic life. Perhaps a few examples will serve to illustrate this complexity.

The Mediterranean and Black Seas, while several thousand feet deep, are so nearly cut off from the ocean by shallow straits that they have little tidal range and a vast amount of relatively stagnant water at deeper levels. The Mediterranean has warmer surface water than the open ocean at the same latitude because of the high amount of sunshine and limited circulation, and a greater surface salinity because of the high evaporation rate and the scarcity of really large rivers other than the Nile flowing into it.

The Caspian Sea, cut off from the Mediterranean-Black Sea by a slight change in the position of the land in recent geologic time, has dwindled in size and now has a surface that is more than 90 feet below sea level. Its northern end has shallow water that is nearly fresh from the inflow of the Volga and which readily freezes in winter, while the deeper, southern part of the Caspian is more like that of the Black Sea.

The Baltic Sea, connected with the ocean only through the narrow water passages of the Great Belt, Little Belt, and Oresund, receives the surface drainage of a large land area, and hence is much less salty than the ocean, especially in its Gulf of Bothnia and Gulf of Finland extremities. The formation of ice in winter interferes seriously with navigation in these gulfs for a period of from two to four months, although the wider part of the Baltic does not freeze.

Along the western side of the Atlantic, the warm Caribbean Sea is fringed by countless islands formed by crustal deformation, volcanic activity, or coral growth. It has a depth of more than 20,000 feet in the Bartlett Trough, and a population which includes large turtles, an abundance of shrimp, and various other crustaceans which thrive in such subtropical waters.

The Sea of Japan is one of four large seas along the eastern margin of Asia; it is separated by arcs of mountainous islands from the open

Pacific. At its southern end in the strategic Korea Strait is Tsushima Island, near which the Japanese navy destroyed a Russian fleet in May, 1905. At its northern end is the narrow Tatar Strait under which a pipeline carries oil from Sakhalin Island to the Russian mainland. While the two straits are shallow, much of the sea is around 10,000 feet deep with a generally mountainous shoreline. Fishing competition is intense between the fishermen of the four adjacent nations, Japan, the Soviet Union, and the two Koreas.

On the thinly populated northern margin of Australia is a square-shaped indentation between 400 and 500 miles on a side called the Gulf of Carpentaria. The gulf is a shallow one, seldom more than 250 feet deep. The slope of the shore is so gradual that high tides, heavy summer rains, and occasional storms cause a shifting beach where salt-tolerant Mitchell grass, palm trees, and mangrove compete for position. It is an area of lonely cattle stations with aboriginal herdsmen and seasonal rivers well-stocked with crocodiles.

The Bering Sea lies between North America and Asia, separated from the Arctic Sea by a 55-mile strait less than 175 feet in depth and from the Pacific Ocean by the rocky and usually fog-bound Aleutian Islands. It is a huge sea of almost 900,000 square miles where the winters are so long and cold that the western shores are ice-bound from October to June. In the Bering Strait the land area of the two big powers is separated by only 2.4 miles of salt water, between the Russian island of Big Diomede and the American island of Little Diomede, a short commuting distance for Eskimos or Chukchis in kayaks who resent political interference with their migrations. The fur of the sea otter was the big attraction for Russian imperialism in these waters in the eighteenth century. Following the reduction of the sea otter by over-fishing and the purchase of Alaska by the United States in 1867, the seal fisheries of the Bering Sea became the object of a major international law case between the United States and the United Kingdom. An international tribunal ruled in 1893 that the United States had no right to exclude British sealers except within the three-mile limit of U.S. territorial waters. Finally, in 1911 pelagic sealing was abolished in the area for all nations in a conservation agreement signed by the governments of the United States, the United Kingdom, Russia, and Japan. Since World War II there have been prolonged negotiations between the governments of the United States, the U.S.S.R., Canada, and Japan over salmon-fishing rights in the area.[18]

Hudson Bay is a vast area (475,000 square miles) of Canadian territorial waters which Henry Hudson, in the early seventeenth century, prodded unsuccessfully in search of a northwest passage to the Orient.

---

[18]Julian V. Minghi, "The Conflict of Salmon Fishing Policies in the North Pacific," *Pacific Viewpoint,* vol. 2, March, 1961, pp. 59-84.

The bay is landlocked except for the usually ice-blocked Hudson Strait which opens eastward to the Atlantic, and the narrow and difficult Fury and Hecla Strait which serves as an outlet from Foxe Basin between the Melville peninsula and Baffin Island. Whaling has been profitable in Hudson Bay, but the short summer shipping season and the high insurance rates have discouraged large-volume wheat shipments from the railroad at Port Churchill, even though the distance through it to Liverpool from Winnipeg is 500 miles less than from Winnipeg via Montreal. Since World War II, several weather and radar stations have been located on and near the shoreline of Hudson Bay.

The Gulf of California, which nearly separates the dry and rocky peninsula of Lower California from the rest of Mexico, occupies a long narrow syncline, a down-fold between two parallel mountain ranges. While over a mile deep in places, the north end of the gulf is shallow from the silt brought in by the great Colorado River. The climate of the shores is desert or near-desert in character, and the aboriginal population lived largely on shellfish. Later, under the Spaniards, a pearl industry of modest importance developed. At the present time, coastal resorts of the Acapulco type, although smaller, are beginning to make their appearance.

The Gulf of Aqaba is a small but now well-known gulf in the Middle East, between the Sinai Peninsula and the northwestern corner of Saudi Arabia. Although about a hundred miles long, it is only 12 to 17 miles wide and much narrower than that at the Strait of Tiran which gives entrance to it. The gulf occupies part of the great rift valley system of eastern Africa and the Middle East. It is more than 600 feet deep almost to its head where the new nations of Jordan and Israel have developed the ports of Aqaba and Elath. The attempt of the United Arab Republic to block Israel's access to Elath after the United Nations police force was withdrawn from Sharm-el Sheik on the western bank of the gulf precipitated the six-day war of early June, 1967.

Economic Significance of the Oceans and Bordering Seas

More than 99 percent of the world's population lives with some permanence on the land and less than one percent on the water. The occupants of houseboats along the margins of the land, of fishing boats, merchant ships, naval vessels, patrol craft, floating scientific stations, and drilling platforms would total, at most, a few million persons. A bare beginning has been made on the design of structures which would permit permanent human settlements on the bottom of the sea, and it can reasonably be presumed that for a long time to come the great majority of mankind will prefer life on land.

Although actually inhabited by only a tiny fraction of mankind, the oceans and bordering seas have significance for man in a wide variety of ways. Space permits only a brief mention of some of the major kinds of ecological relationships between man and the sea.

Since earliest times, one of the major concerns of organized societies has been self-defense. Island or peninsular locations have facilitated the protection of numerous groups from conquest by or submergence within adjacent larger groups. The primitive inhabitants of the Andaman Islands, the Polynesians of Samoa and the Tonga Islands, the Koreans, the Japanese, and the British are examples of population groups whose cultural identity has been kept intact to a considerable degree with the help of isolation by water. Modern industrial nations commonly prepare elaborate fixed coastal defenses and equip themselves with mine layers, submarines, destroyers, aircraft carriers, and a wide range of other equipment to ward off potential threats from the sea.

No one knows when the first trading expedition by sea was launched, but the epic poems attributed to Homer and the archeologically documented commerce of the early Minoan, Egyptian, and Phoenician civilizations testify to its antiquity. Scheduled and unscheduled merchant vessels now criss-cross the oceans and bordering seas in a commercial network of amazing intricacy. Under the flags of the United States, the United Kingdom, Liberia, Panama, the Soviet Union, Norway, Japan, and many other countries, hundreds of tankers carry petroleum and thousands of merchant vessels carry less specialized cargo by sea. Passenger vessels are far fewer in number than freighters, but many tourists and business travelers still prefer the restful luxury of the ocean liners to the speed of the international airlines. Since the successful completion of the first trans-Atlantic cable in 1866, underwater telegraphic and telephonic communication has become commonplace, and radio-telephone service from ships at sea is widely available.

Commercial fishing in the oceans and seas is a way of life and a means of livelihood for several million people. The product of their effort is important to hundreds of millions on the land. While the search for whales, prized especially for their oil, began to shift from the northern seas to the cold waters around Antarctica about the beginning of this century, the most important commercial fisheries remain in the northern hemisphere. Partly because there seems to be a greater abundance of plankton and, hence, of fish, but also because there are enormous markets near at hand, the three major areas of commercial fishing are: (a) the waters off eastern Asia, around Japan, Korea, southeastern Siberia, China, and the Philippines; (b) off northwestern Europe in the North Sea, the Norwegian Sea, the Bay of Biscay, and the waters around Iceland; and (c) off eastern North America in the waters around Newfoundland, Nova Scotia, and the southern tip of Greenland.

While some of the sea animals of commercial interest are attached to the shallow sea floor, like the oyster, or crawl around slowly on it, like the lobster or the king crab, most of them swim freely through the water, singly or in schools, at a favorite depth. *Demersal fish,* such as cod, haddock, hake, plaice, sole, and halibut, live near the bottom of the water on the continental shelves. *Pelagic fish* live on or near the surface. These include eel, mackerel, the herring family (sprat, pilchard, sardines, etc.) and, in warmer waters, tunny, bonitoes, and sharks. The *benthic,* or bottom-dwelling varieties are usually caught or cut free by hand or captured in simple traps. Bringing in the free-swimming species requires a variety of techniques. Demersal fishing involves the use of weighted lines or of some type of trawling operation in which a net is dragged near or directly on the sea floor and is raised periodically with a steam or electric-powered winch. Pelagic fish are taken in from lines with baited hooks at intervals along the line or from purse seines, or other types of seines, with the net supported by floating buoys or directly by the boat or boats which hold it close to the surface.[19]

With the increasingly rapid utilization of the mineral resources of the land, man is giving much more attention than ever before in history to the minerals in sea water. At least fifty of the known elements have been found to be present, but most of them are there only in extremely minute quantities. Five salts constitute the great bulk of the mineral content in sea water: sodium chloride (77.8 percent of the total salts), magnesium chloride, magnesium sulfate, calcium sulfate, and potassium sulfate. Of these, sodium chloride, common salt ($NaCl$), has been recovered commercially by a simple evaporation process from shallow beds or "salt pans" at numerous points along the coast since ancient times. Along with the mining of crystallized "rock salt" and the recovery of salt from pumped brines, it is still one of the important sources of marketable salt. Since shortly before World War II, magnesium chloride brines from the Gulf of Mexico have been processed commercially into magnesium metal, which is much in demand for its happy combination of strength and very light weight. Bromine is another mineral which is recovered commercially from sea water. At a number of places in the world, including the Dead Sea on the Israel-Jordan border and Searles Lake in California, potash is recovered from brine lakes on land where the concentration of salts is greater than in the open oceans. It is not at present economically feasible to extract the very substantial amounts of gold and silver which are said to be present in each cubic mile of ocean water.[20] Although

[19]King, *op. cit.,* pp. 272-296.

[20]Rachel L. Carson, *The Sea Around Us,* New York: The New American Library, Inc. (Signet Science series), rev. ed., 1961, p. 176.

not exactly in the same relationship to sea water as the minerals just mentioned, it is common to include in discussions of mineral resources of the sea the vast wealth of oil present in the rocks of the submerged continental shelf. This oil, along with the associated natural gas, is under active development in the territorial waters of many countries.

The recreational value of the oceans and their shores is evident in the growth of hundreds of cities and towns whose principal function is to provide lodging, food, and pleasurable diversions for their vacationing residents. Some of these coastal resort centers have grown into sizable cities as, for instance, Atlantic City and Miami Beach in the United States, Blackpool and Brighton in England, Nice and Biarritz in France, and Yalta and Sochi in the Soviet Union. Bathing, water skiing, scuba diving, yachting, deep-sea fishing, and inter-island luxury cruises are increasingly popular diversions for urban populations with money to spend and vacation time in which to spend it.

FIGURE 2.2 A Bathing Beach on Cape Cod. Courtesy National Park Service. The water may be cold, but the sand is warm and the salt air is refreshing on the Cape Cod National Seashore Beach near Eastham, Massachusetts.

Political aspects of the oceans and bordering seas are of growing concern and are the subject of books, government documents, and numerous journal articles. Space is available here only to identify some of the categories of interest and concern. Control of the seas, in the sense of being able to dominate ocean-going commerce or assert naval superiority, has been a recurring objective of political authorities almost since the dawn of history. Ancient Phoenicia and Athens, medieval Genoa and Venice, Spain and England in the early modern period, and most of the major powers of the twentieth century have given a high priority to sea power. Not only navies but merchant fleets as well have received support and subsidies from public authorities. Since the time of Hugo Grotius (1583-1645), the Dutch scholar often considered to be the "father of international law," a body of national and international statutes and precedents has developed concerning the legal rights and privileges of captains and seamen, the status of belligerent and neutral vessels in time of war, the control of piracy and smuggling, the rights to shelter in time of storm and salvage after a shipwreck, or to fish or prevent fishing.

In the crowded, competitive, and at times hostile kind of world that exists as this book is being written, nations are much concerned about such matters as the extent of offshore territorial jurisdiction, the pollution of coastal waters, mineral resources of the continental shelf, exclusive fishing zones, the operation of "spy ships" and illegal radio transmitters, atomic testing over international waters, and scientific stations on the bottom of the ocean. International conferences on the "law of the sea" at Geneva in 1958 and 1960 brought together legal experts and other interested parties from nearly a hundred countries. A few decisions were reached, and many issues were sharpened for consideration at subsequent international conferences.

Chapter 3

Water in the Atmosphere

Water Vapor and Humidity

While water is most commonly in liquid form and is ordinarily considered a liquid, its existence as a gas called *water vapor* is of extraordinary importance. The presence of a small amount of water vapor in the atmosphere, never amounting to more than about five percent of the total volume, is what keeps diurnal (night to day) temperature fluctuations within tolerable limits and prevents a person's skin from wrinkling like that of a crocodile or a sun-dried prune. Perhaps even more important is the fact that only in the form of water vapor are natural processes able to take fresh water from the ocean and move it onto the land, even to the top of the highest mountains, so as to maintain the hydrologic cycle on which people are so totally dependent for existence.

As noted in Chapter 1, some 280 cubic miles of water is evaporated each day from the water and land surfaces of the earth, more than four-fifths of this from the oceans. Heat from the sun is the motivating force which makes this transfer of moisture into the atmosphere possible. At its average distance from the sun of about 93,000,000 miles, the earth intercepts less than 1/2,000,000,000 of the total outflow of solar energy, but even this minuscule fraction amounts to 23,000,000,000,000 horsepower of energy.[1] The energy in falling water, in wind, and in thunderstorms reflects this same solar energy in different forms.

Of the *solar radiation* received by the earth and its atmosphere, about one-fourth (24 percent) is intercepted by clouds and reflected directly back into space. Another fraction, about six percent, is similarly reflected back into space by air molecules, atmospheric dust, and

[1]Glenn T. Trewartha, *An Introduction to Climate,* 3d ed., New York: McGraw-Hill, Inc., 1954, p. 7.

water vapor. Some 22 percent of the incident radiation is reflected downward by the atmosphere and its constituents, reaching the surface as diffuse sky radiation. Approximately 17 percent is absorbed by atmospheric constituents and clouds, leaving 31 percent to reach the surface as direct-beam solar radiation. Altogether, a little more than half of the incoming solar radiation finds its way to the earth's surface. Because of its somewhat drier atmosphere, the northern hemisphere receives a little more radiation at the surface than the southern hemisphere. The maximum radiation at the surface occurs in the large deserts of the northern hemisphere, in northern Africa, southwestern Asia, and the southwestern interior of North America, where it is three to four times that received by the surface of the northern Atlantic or northern Pacific Ocean.[2]

The amount of water vapor actually present in the air in a specific location is a function of the wetness or dryness of the adjacent surface of the earth, the mixing effect of the wind and air currents, and, especially, of the temperature. It is not easy to get precise information as to the actual amount of water loss by *evaporation* from the surface of the ocean, from lakes or reservoirs, or from damp ground. With the same temperature and wind conditions, the loss is greater per unit area from small water surfaces than from large ones. For more than half a century the United States Weather Bureau has been making comparative evaporation studies with a standardized pan about four feet in diameter and ten inches deep. It has tried to compare the water lost from the pans with that from adjacent free water surfaces. An annual evaporation loss of more than ten feet has been noted from pans set in desert environments (near Lake Mead, Salton Sea, and others), and a loss of five to ten feet from many larger lakes and reservoirs seems clearly established.[3] For purposes of climatic classification and also with a view to better understanding the earth's heat balance, Professor C. Warren Thornthwaite and others have found the concept of *potential evapo-transpiration* a useful one.[4]

How much water is held in the air as vapor over a given place at a given time cannot be known with great precision and, of course, it is changing constantly. If all the water vapor in the earth's atmosphere were somehow condensed and precipitated at the same time, it is thought that it would form a layer less than one inch thick.[5] The water vapor

[2]William D. Sellers, *Physical Climatology,* Chicago: University of Chicago Press, 1965, pp. 19-25.

[3]*Ibid.*, pp. 150-156.

[4]C. Warren Thornthwaite, "An Approach Toward a Rational Classification of Climate," *Geographical Review*, vol. 38, January, 1948, pp. 55-94.

[5]P. H. Kuenen, *Realms of Water: Some Aspects of Its Cycle in Nature*, New York: John Wiley & Sons, Inc., (Science Editions), 1963, p. 89. (Originally published in Dutch, 1955).

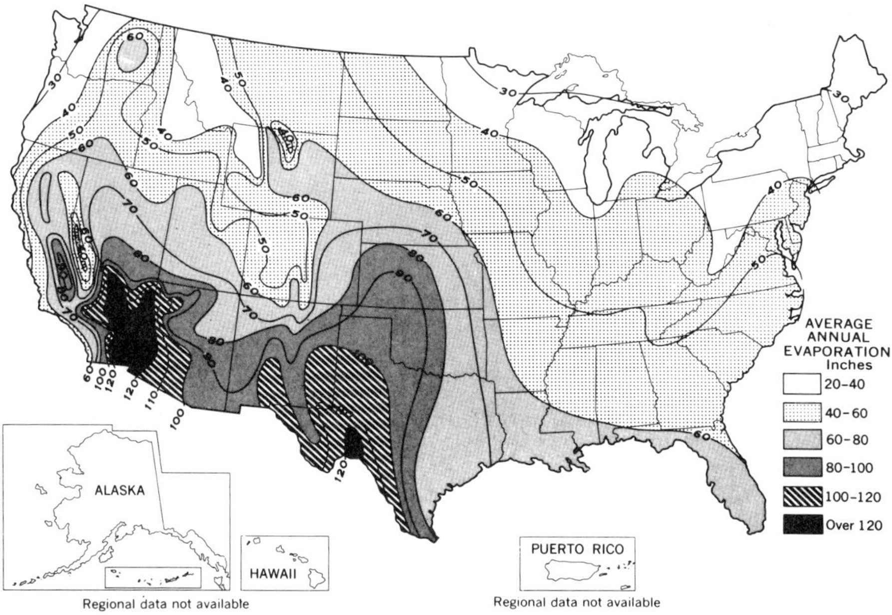

FIGURE 3.1 Average Annual Lake Evaporation in the United States. Over most of the country the actual or potential evaporation from free water surfaces exceeds precipitation. From United States Water Resources Council, **The Nation's Water Resources,** p. 3-2-3.

tends to remain near the surface of the earth from which it comes. Almost half of it is normally present within a mile of the earth's surface, and there is very little vapor detectable in the rarified atmosphere six or seven miles above the surface.

Just as air exerts pressure that is measureable by a barometer, the water vapor within the air exerts its own pressure. *Vapor pressure* reaches its maximum level when the air is saturated—that is, when it can hold no more at a given temperature without condensation. The condition of the atmosphere in relation to its water vapor content is referred to as *humidity*. It can be measured as *specific humidity* (the weight of water vapor per unit weight of air, usually expressed in grams of vapor per kilogram of natural air), as *absolute humidity* (the weight of water vapor per unit volume of air as, for instance, in grains or grams per cubic foot or cubic meter), or as *relative humidity* (the percentage ratio between the amount of water vapor actually present and the maximum amount which could be present under the existing conditions of pressure and temperature). Most people are familiar with expressions of relative humidity as read in percent on the home hygrometer or as reported in the various news media.

The increased ability of the atmosphere to support water vapor as the temperature increases is impressive, as shown in TABLE 3.1.

TABLE 3.1

Water Vapor Present in One Cubic Meter of Saturated Air

Temperature in Fahrenheit Degrees	Moisture Content in Grams
–40	.085
–20	.33
0	1.0
32	4.87
60	12.91
80	25.0
100	50.0

*Adapted from a table in George R. Rumney, *Climatology and the World's Climates*, New York: The Macmillan Company, 1968, p. 10.

From TABLE 3.1 it is evident that air at 40 degrees below zero can hold only one-twelfth as much moisture as air at zero degrees Fahrenheit and that raising the temperature of air from 32 degrees above zero, the freezing point of fresh water, to 80 degrees, increases its vapor-holding capacity by more than five times. While it is not exactly correct to say "it is too cold to snow" or "it is too cold for a thunderstorm," there is a substantial measure of truth in such expressions. In a typical daily situation within a given air mass, the relative humidity is generally lowest during the warmest part of the afternoon and highest during the night time and early morning hours, when fog is most likely to form or dew to appear on the ground in response to local cooling beyond the level of maximum vapor pressure at the prevailing temperature.

Condensation is the conversion of water vapor to water droplets, either in the atmosphere as *fog* or *clouds* or on the ground as *dew*, when the atmosphere is cooled to the level of *saturation*. This level is also known as the *dew point*. Cooling of the type which may lead to condensation can be the result of (a) surface heat loss due to nighttime *radiation*, (b) ascending air currents as in *convectional circulation*, (c) the movement of air up a mountain slope in response to surface pressure differences, which leads to *orographic cooling*, (d) direct contact with a cooler air mass or a cooler ground surface in which case there is heat loss through *conduction*, or (e) expansion with upward movement along a sloping *front*. When two air masses of differing temperature and moisture characteristics confront each other, as they do in many places on the globe, the lighter one is commonly forced upward and cooling occurs along the zone of contact. Weather forecasting is the meteorologist's effort to predict the behavior of adjacent air masses and to suggest the consequences in terms of temperature change, wind conditions, cloudiness, precipitation, and storm development.

When air is cooled to the dew point, or approximately that level, condensation of some of the water vapor present takes place. If the dew point is reached when the temperature is below 32 degrees Fahrenheit (0 degrees Centigrade) the process of *sublimation* may occur. This is a direct transfer from the water vapor state to the solid state in the form of white frost, snow flakes, or ice crystals. Condensation, it appears, can only take place when solid particles are present to serve as nuclei, but there is usually an ample supply of such microscopic particles of salt, smoke, or dust present, even at higher levels of the atmosphere. The most strongly hygroscopic of these particles, such as salt blown inland from ocean spray or the smoke which results from the combustion of coal or oil, may stimulate some condensation even at relative humidities well below the dew point, evident in the form of light *haze*. When the extent of condensation is greater, the usual result is the formation of fog or, if it occurs above the ground surface, of clouds. A certain amount of haze is a common daytime occurrence in humid regions, especially in the summer season. Fog may occur as many as 80 or more days a year along some coasts in high latitudes, or in middle or even low latitudes where winds blow over a cool ocean current.

Cloudiness is a common condition of the atmosphere in most parts of the earth. This is a matter of simple observation for those who do much traveling by air or who have examined photographs of the earth's surface taken from satellites. The percentage of the sky covered by clouds averages quite low in lower middle latitudes, say from 20 to 35 degrees on either side of the equator. Nearer the equator and in higher latitudes, mean cloudiness commonly exceeds 50 percent and may be much higher. Seasonal and even diurnal variations of cloudiness are characteristic of many areas.

Cloud formation occurs at various levels of the atmosphere, but it is not common more than six or seven miles above the surface. At this level, when clouds are present they are usually of the thin feathery type known as *cirrus,* formed entirely from ice crystals. Other forms of clouds, such as the billowy *cumulous* or even-gray *stratus* types, form at lower levels. Names have been given to many intermediate types or variations of the basic cirrus, cumulous, and stratus, such as *cirrostratus, altocumulus,* and *nimbostratus*. Cloud formations are constantly changing, and many individuals find as much pleasure in cloud-watching as others find in bird-watching or girl-watching.

Precipitation occurs when condensed or sublimated moisture in the air reaches such a density that rising air currents can no longer keep it all aloft. It may reach the surface in a variety of forms, depending upon its experience in descent, for instance, as *drizzle, rain, sleet, snow,* or *hail.* Space is not available here to discuss all of the processes at work in shaping the various forms of precipitation, but such general explanations are available in standard textbooks of *meteorology* and *climatology.* As mentioned earlier, some sequence of atmospheric cooling is involved in the condensation-precipitation process. And it is common to classify types of precipitation in relation to the predominant means of cooling involved, convectional, frontal, or orographic.

Convectional precipitation is that which occurs as a result of the warming of surface air to the point where it is both warmer and lighter than the air which is around it and above it. The updraft of air set in motion by the surface heating is maintained for a period of time by the generally horizontal inflow of air toward the spot or line of maximum heating. As the updraft continues, the air within it cools by expansion until it reaches a level in equilibrium with the surrounding air at the same altitude. Cooling initially takes place at a rate of about 5.5 degrees Fahrenheit for each 1,000 feet of rise. This is the so-called *dry lapse rate.* If during the uplift process the air is cooled to the dew point and rapid condensation sets in, clouds form and thicken, and precipitation may result. The straight-line bottom of cumulous clouds floating over the earth on a hot summer afternoon reflects the level at which rapid condensation begins to take place. When the air continues to rise after heavy condensation sets in, it cools at the much slower *wet lapse rate,* about 2.7 degrees per 1,000 feet. The slower rate of cooling is the result of the release of heat, known as the *latent heat of condensation.* This release of heat tends to maintain the lightness of the air and facilitate its further rise. Sometimes the updrafts of air occur with great violence and present a threat to the safety of aircraft and certainly to the comfort of passengers flying through them.

The cooling of air by the convectional process, leading to frequent afternoon and early evening rains all or most of the year, is typical in equatorial regions. Convection on a smaller scale, often leading to local thundershowers, is commonplace in interior regions of middle latitudes. The convectional process, or something very like it, also takes place in zones of unstable air along the margin of contrasting air masses in either low or middle latitudes.

A second general type of precipitation is *frontal or convergent,* the type which occurs along the zone of confrontation of two air masses with different density, temperature, and water vapor characteristics. The heavier, cooler air mass typically hugs the ground, forcing the lighter, warmer one with a higher vapor content to override it, ascending a gentle, oblique slope. During the ascent, cooling of the lighter air mass results from both expansion and conduction, bringing on condensation and, frequently, precipitation. The actual line of contact of the two air masses on the ground may be moving either north or south—hence, the terms warm front and cold front. The cooling process is similar in either case. Since frontal cooling is more gradual than that in the vertical ascent of air associated with convection, frontal rains are usually more gentle in nature and may continue at a given location for several hours or even days. Frontal precipitation may occur at most any latitude at any season, but it is especially characteristic of cooler seasons in the interior plains of middle latitudes. The term *cyclonic precipitation* has often been used for that which results from atmospheric convergence in the moving low pressure cells which form along the general line of a major front.

The third type of precipitation to be considered here, *orographic,* comes with the cooling of air that is drawn, in response to surface pressure contrasts, up a mountain slope. Mountains along or near the coast on the windward side of continents may receive very heavy precipitation in the form of rain and/or snow on the side facing the ocean. The windward side of a land mass, either a continent or an island, is ordinarily the east side in the latitudes where the *trade winds* blow, between 10° and 30° N. and in the comparable latitudinal belt in the southern hemisphere. In middle and upper-middle latitudes, especially between about 40° and 60° in either hemisphere, the windward side is the west side. These are the zones of the so-called *prevailing westerlies.*

A cursory examination of the world precipitation map reveals that precipitation is distributed very unequally over the earth. The areas of lowest precipitation, under ten inches per year, can be explained in relatively simple terms. Some of these areas are in polar and subpolar regions where the air is nearly always cool to cold and so cannot retain much moisture in the form of water vapor. When there is not much vapor in the air, even though the relative humidity may be high, not much can be released with the frontal cooling that does occur in these high latitudes. In middle and lower latitudes, the areas of very low rainfall tend to be on the lee side of mountains, along certain coasts paralleled by cold ocean currents, and in the deep interior of mid-latitude continents far from the oceanic sources of moisture.

Areas of highest rainfall, over 60 inches a year, are generally in one of two types of locations, either near the equator where there is a more or less continual process of convectional cooling going on, or along windward coasts where the steady trade winds or the less-steady

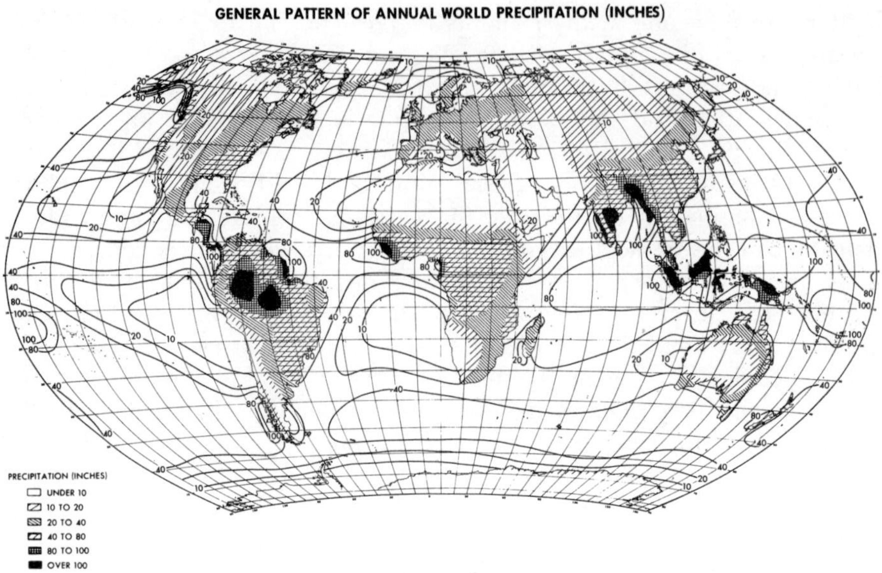

FIGURE 3.2 General Pattern of Annual World Precipitation (Inches). This map shows the general distribution of precipitation over both land and water surfaces. It is not possible on a map of this scale to show the extreme variability of precipitation which occurs within short horizontal distances on land, especially in mountain regions. From ESSA, U.S. Department of Commerce, **Climates of the World,** p. 4.

"prevailing" westerlies bring a fresh supply of vapor-laden air. This air is cooled on contact with the land or by orographic lifting on the sea-facing slope of nearby highlands. Fortunately for the habitability of large areas of interior plains in middle latitudes as in much of interior North America or Europe north of the Alps, surface pressure conditions develop which draw in moisture from source areas hundreds of miles away. The convergence of air masses in the deep interior of mid-latitude continents may supply from 15 to 40 or more inches of frontal and convectional precipitation, mostly in the warmer half of the year.

Only rarely does the precipitation which falls at a particular spot on the earth's surface arrive with reasonably equal distribution during the course of the year. In desert and semiarid regions the moisture supply is notably unreliable, and as much may arrive in one or two months some years as arrives in all twelve months during other years. It is not especially unusual for a desert settlement to receive most of its annual rainfall for a year in one heavy shower. In tropical and subtropical areas the season of maximum rainfall is likely to accompany or follow shortly after the high-sun period—that is, a period of maximum *solar insolation.* In the interior of the mid-latitude continents, the season of heaviest precipitation also tends to be in the high-sun period, especially in late spring and early summer. Winter or low-sun precipitation predominates along certain west-facing coasts in upper middle latitudes and in regions of so-called Mediterranean climate on the west side of large land masses in a zone between about the 30th and 40th parallel in either hemisphere.

A somewhat special type of cyclical rainfall distribution occurs in southern and eastern Asia in association with the seasonal reversal of wind direction known as the *monsoon.* The summer low pressure center which develops over the northern plain of India and Pakistan, and eventually extends into the deep interior of Asia, draws in warm moist air from over the Indian Ocean and the western Pacific, bringing heavy, late spring and summer rains to the southeastern and eastern portions of the continent and to nearby islands. The outblowing winds of winter, fed by the very cold, heavy air which develops over the continental interior at that season, are responsible for several months of low precipitation in the same regions.

The nature, amount, and cyclical distribution of precipitation over the surface of the earth are of concern to man for a great variety of reasons. Must he build shelter against blizzards and provide himself with skis and snowplows? Does it rain enough to raise food crops without irrigation? Will the valley flood and the bridge wash out again next year? What can be done to reduce soil erosion on these hillsides? Will the city have an adequate water supply without drilling some more wells or constructing that reservoir? Is it safe to schedule the company picnic on Saturday afternoon? These and thousands of similar questions are confronted by individuals, groups of persons, and public authorities somewhere every day of the year.

Chapter 4

Snow and Ice

Familiar forms of water in the solid state are snow and ice. Snow, which can be formed only in the atmosphere, begins with the sublimation of water vapor at temperatures below freezing into tiny particles of ice. As these ice particles are growing they assume platy or prismatic hexagonal forms of great variety and marvelous beauty. In their movement, stirred by the wind, they hook onto each other and coalesce into what we call snowflakes. Most snowflakes are very short-lived, for when they reach a weight sufficient to descend toward the earth their usual fate is to melt and fall to lower levels as rain, and then perhaps to evaporate before they reach the surface. Some snow, however, reaches the ground where it may melt at once or it may survive for a few weeks or months or possibly become compacted and gradually transformed into ice.

Apart from its role in snow formation, ice appears in the atmosphere as sleet or hail, which can both be defined as frozen rain. Sleet is a mixture of fine rain and ice which will adhere to such objects as trees, fences, bridges, flying aircraft, and automobiles. Hailstones are larger balls of ice, often very damaging to crops, formed of frozen raindrops which have grown with concentric additions as strong summer convective currents carry them upward again and again before they finally grow large and heavy enough to reach the ground.

Ice forms under appropriate temperature conditions at the surface of both fresh and brackish water bodies and develops underground by the freezing of soil moisture or ground water. In high latitudes permanently frozen ground, or *permafrost,* is a common phenomenon and may extend to considerable depths.

Distribution of Snow

Snow has been observed over about 80 percent of the land surface of the globe, and a more or less permanent ice cover occupies approximately 11 percent of that surface. Depending upon the season, from 30 to 50 percent of the surface is normally covered with ice and/or snow. These figures, however, include areas where the snow cover is present for only a brief period. Snow is an important phenomenon, remaining three to four months over about a third of the earth's surface and at least six months over 20 to 25 percent of it.[1] It is a major influence on plant and animal life, on human mobility, and on many economic and social considerations including spring floods, summer water supplies, and winter sports.

While snow falls at all latitudes in the high mountains, it is quite rare at sea level and on low plains between 30 degrees north and 30 degrees south latitude. Because of the moderating influence of winds blowing off the ocean, snow does not appear regularly as far south

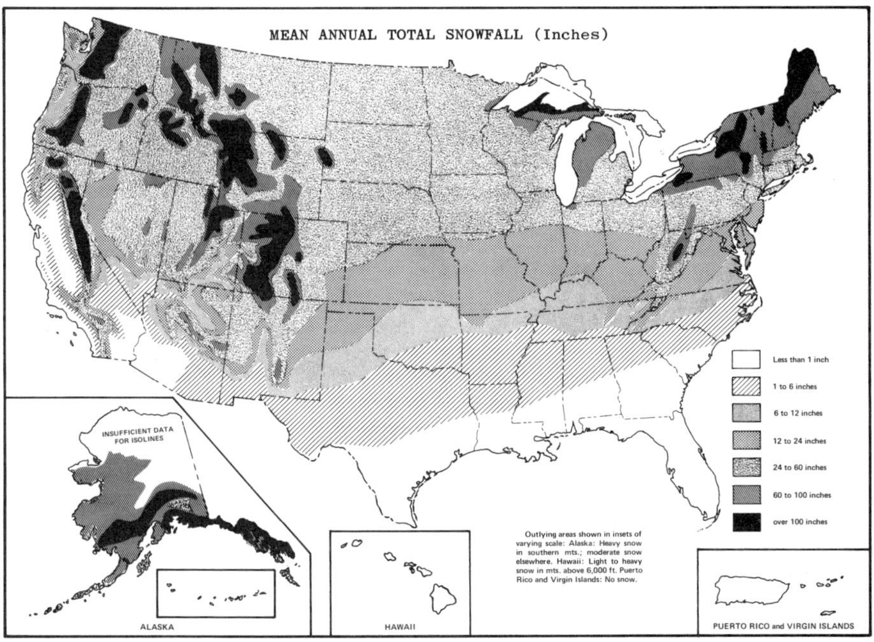

FIGURE 4.1 Average Annual Snowfall in the United States (Inches). Courtesy Environmental Science Services Administration (redrawn and simplified). The amount of snowfall is a function of both latitude and altitude.

[1]Jean Corbel, *Neiges et glaciers,* Paris: Librairie Armand Colin (Collection Armand Colin, No. 361, Section de Geographie), 1962, p. 7.

on the western margin of North America or Eurasia as it does on the eastern side of those continents. Light snow occasionally falls on the northern and southern extremities of Africa, even on the lowlands, and the narrow southern end of South America experiences heavy snowfall, even on the Patagonian plateau east of the Andes. Seasonal snow is very widespread on the northern two-thirds of Asia and over nearly all of Europe, the United States, and Canada. Snow is fairly unusual at Athens and New Orleans, and it ordinarily stays on the ground for only brief periods at Genoa, Marseilles, Brest, Liverpool, Los Angeles, and San Francisco.

The duration of the snow cover generally increases poleward and inland from areas of marginal occurrence. While Tucson, Arizona, experiences snow, on the average, only two days and Oklahoma City about two weeks per year, the snow blanket remains on the plains of Nebraska and Iowa for two or three months and along the United States-Canadian border for more than four months. In the Paris Basin or around the southern shores of the North Sea the snowfall is light, and there is usually rapid melting within a day or two after each occurrence. On the Ardennes Plateau, however, within about a hundred miles of the North Sea coast, the snow remains much longer, and it forms a continuous cover for many weeks on the plain of Alsace in eastern France, on the Swiss Plateau, and on the extensive lowlands of Germany, Poland, and the U.S.S.R. Snow persists on the ground at Moscow, on the average, for 140 days, and at Vorkuta, an important coal-mining town near the Arctic Circle in European Russia, for 240 days. There is snow on the ground for seven months at Schefferville in eastern Canada, where a great iron mining industry has developed since World War II.[2]

While the instrumentation for measuring snowfall is somewhat less accurate than that used for rainfall, reasonably good data is available concerning average snowfall over the United States and Canada, Europe and the Soviet Union, Australia, and New Zealand. Only about two inches of snow is to be anticipated in an average winter in central Texas or Central Mississippi, while about ten inches can be expected at Nashville, Tennessee, or Norfolk, Virginia. Twenty inches is about normal in northwestern Kansas and central Illinois, and thirty inches or more comes regularly to eastern Montana, western Minnesota, northern Ohio, and southern New England. Even on the plains, some communities receive eighty inches or more each winter if they happen to be situated along the south shore of Lake Superior or Lake Ontario.[3]

---

[2]*Ibid.*, p. 24.

[3]Map of average annual distribution of snow over the United States in Corydon Bell, *The Wonder of Snow*, New York: Hill and Wang, Inc., 1957, p. 12.

The amount of snowfall is less in regions of extremely cold winters than in moderately cold areas. The bitter cold interior of northeastern Siberia receives less snowfall than the much less cold plain between Moscow and Leningrad. Additional snow falls frequently on Antarctica, Greenland, and other ice-capped areas of the polar and subpolar regions—several feet per year in some places, it appears—but the relatively short period of record keeping and the sparsity of the weather station network in these areas make it difficult to assemble very complete information.

Snowfall is sometimes extremely heavy in very mountainous regions, especially where high ranges parallel a coast with characteristically onshore winds. In western North America the Pacific slopes of the British Columbia coast ranges, the Cascade Mountains of Washington and Oregon, and the Sierra Nevada of California illustrate the point. Between elevations of 6,000 and 8,000 feet in the Sierra Nevada and between 4,000 and 6,000 feet in the Cascades as well as in their continuation in Canada, the normal winter snowfall is from 300 to 600 inches, about 25 to 50 feet. Surely in competition for the distinction of being the snowiest place on earth are such weather stations as the one at Tamarack, California (elevation, 8,000 feet) which in the winter of 1906-07 measured 884 inches of snow and the Mount Rainier-Paradise Ranger Station (elevation, 5,500 feet) which in 1955-56 recorded an even 1,000 inches, more than 80 feet![4] Extremely heavy snows have also been measured in the Alps and on some of the mountains of Japan, central Asia, and New Zealand.

Snow slides, called *avalanches,* are a threat to hikers and winter motorists in many mountain regions, especially when there is a fresh fall of snow on a slope of 20 degrees or more where a previous thaw coated the underlying snow surface with ice. Numerous travelers have been trapped on mountain roads and trails by storms which may add from five to eight feet thick of fresh snow in a single 24-hour period. Each winter the news media have occasion to report disasters affecting livestock, motorists, and even passenger trains stranded by heavy snows. In January, 1952, the Southern Pacific's passenger streamliner, the *City of San Francisco,* was trapped by snow for four days on the west slope of the Sierra Nevada below Donner Pass, where about 35 persons in George Donner's emigrant train were trapped and died in the deep snows in the winter of 1846-47.[5] The worst avalanche disaster of recent years was that which occurred on January 19, 1962, when a mass of snow and ice estimated at 2,800,000 cubic yards broke loose on the

[4]U.S. Weather Bureau figures as cited in Rumney, *Climatology and the World's Climates, op. cit.,* p. 612.

[5]James L. Dyson, *The World of Ice,* New York: Alfred A. Knopf, Inc., 1962, p. 20.

slopes of Mount Huascaran in the Peruvian Andes, descended on the village of Ranrahirca, and killed about 2,000 men, women, and children. Kuenen reports that on the World War I battle front in the Tyrolese Alps, more than 6,000 Austrian soldiers lost their lives on December 16, 1916, in a series of gigantic avalanches.[6]

At higher elevations in the mountains of middle latitudes, snow may occur at any time of the year. The writer has experienced fresh snow in August in the Colorado Rockies on the Leadville-Aspen road at around 9,000 feet above sea level and at lower elevations in the same month in the mountains of Wyoming and Montana. Snowfalls of one to three inches were recorded on the top of Mount Washington in New Hampshire (elevation, 6,290 feet) in August, 1964, and July, 1965.[7] Near the equator, where there is relatively little seasonal temperature change, occasional snow has been observed as low as 12,000 feet elevation, but the normal, yearly occurrence of snow is restricted to elevations above 16,000 or even 18,000 feet.[8]

The *snow line* is the lower limit of permanent snow cover, which varies in elevation from sea level at the poles to about 20,000 feet at the equator. In southern Greenland it is at about 2,000 feet, in Norway at 4,000 to 5,000 feet, and in the Alps at about 9,000 feet. The direction of the prevailing wind and the slope in reference to the sun both affect the position of the snow line. The snow barely melts by mid-summer in Tuckerman Ravine on Mount Washington, and in some of the north-facing canyons of Ben Nevis (elevation 4,406 feet) in Scotland it melts completely only when there is an exceptionally warm August. In the Himalayas the permanent snow line on the drier north slope is at about 20,000 feet, while on the south slope, where the summer monsoon brings very heavy snow above the zone of heavy rains, it is nearer 16,000 feet. The great volcanic cones of southern Mexico, Mount Popocatepetl (17,887 feet) and Mount Ixtaccihuatl (17,343 feet) are white-capped all year long, as are Mount Chimborazo (20,577 feet) in Ecuador and Mount Kilimanjaro (19,565 feet) in Tanzania, the last two within 200 miles of the equator.[9]

The moisture content of fresh snow varies enormously. The heavy wet snow which weighs down the branches of trees and makes good snowball throwing when the temperature is near the freezing point may have five times as much moisture in it as the equivalent volume of loose, dry snow of the type which falls in the high mountains or on the polar icecaps. In considering the rainfall-equivalent value of

---

[6]Kuenen, *op. cit.*, p. 136.
[7]Rumney, *op. cit.*, p. 611.
[8]*Ibid.*
[9]F. J. Monkhouse, *Principles of Physical Geography*, London: University of London Press, 6th ed., 1965, pp. 230-231.

snow, it is customary to regard ten inches of average snow as equal to one inch of precipitation.

Ice on the Land

Soon after it falls, snow begins to settle and become more compact. If there is surface melting, water percolates into the pore spaces, gradually giving the mass a whitish, granular texture. A certain amount of sublimation also takes place, with water vapor escaping from the snow flakes and re-collecting within the mass. When the pore spaces between the grains of snow are largely sealed off and the density reaches a little more than half that of water, the term applied to the mass is *névé* (French) or *firn* (German). An accumulation of annual layers of compacted snow in locations where the summer melting is not equivalent to the seasonal snowfall is often referred to by glaciologists as a *firn field,* and the place of accumulation as a *firn basin.* When the firn is further compacted to a density of 0.8 to 0.9 it has become *glacial ice,* approaching the 0.917 density of pure ice. As soon as the glacial ice becomes heavy enough to spread out by its own weight and begin to move downhill it has become a glacier.

Formation of ice on rivers and lakes is a natural accompaniment of prolonged sub-freezing temperatures. Since water has a rather high specific heat and is thus slow to change temperature, and in addition is often in motion and therefore mixing, there is a significant time lag between the arrival of sub-freezing air temperatures and the formation of ice. Even in what appears to be still water, as the top layer cools to its maximum density, about 39 degrees Fahrenheit, the surface water sinks to lower levels and is replaced by new surface water that is less cold. When the whole mass has cooled to 39°, such convectional circulation ceases, and further exposure to sub-freezing air will permit ice to form at the surface. Usually ice makes its first appearance along and attached to the river bank or lake shore, although on a clear night with no wind a skim of ice may form over the entire surface of a sizable lake. As winter conditions persist, the layer of ice thickens. Eventually, if it remains cold enough for a sufficiently long period of time, a river may freeze all the way to the bottom of its deepest channel and cease to flow. The low conductivity of ice prevents the indefinite thickening of the frozen layer, however, and an ice depth of greater than eight to ten feet in high-latitude lakes or even in the polar ice pack is unusual.

The action of wind and waves on a layer of ice tends to crack it and break loose segments of it. In the pack ice of Arctic and Antarctic waters, large *floes* may move against each other, crumpling their margins and creating pressure ridges which stand ten or twenty feet above the general level. The same process is responsible for the uneven undersurface of the ice observed by submarines moving beneath it. When ice floes in a lake are separated by open water, the refreezing of this water enlarges the whole mass and may lead to the sudden creation of irregular ridges of ice far out from the shore. Very frequently the ice also presses with great force on the shore, pushing up ridges of rock material several feet thick.

A layer of ice, especially fresh-water ice, is remarkably strong and may be able to support heavy weights. Skaters venture onto ice, sometimes to their peril, when it is only a few inches thick. When the ice on the lakes of central Minnesota freezes to a thickness of two feet or so, it is relatively common to see automobiles and pick-up trucks parked on it far from shore, with the drivers sitting alongside of them in heated tents each fishing through a small specially-chopped hole. During the siege of Leningrad in World War II, when German forces controlled the access highways, relief columns of heavy trucks reached the beleagered city by making night runs across the ice of Lake Ladoga. When the Trans-Siberian Railroad was completed across southern Siberia, the train ferries, which for a time carried the trains across Lake Baikal in the summer season, were replaced in winter by a railroad track laid directly on the ice. Heavy caterpillar-type tractors have been used on the *shelf ice* attached to the shores of northern Canada and Antarctica, and semi-permanent airfields have been developed on large *ice islands* off the Arctic coast.

When the spring thaw finally comes in regions of severe winter cold, the ice cover on the rivers breaks up with great suddenness. Huge blocks of floating ice jostle their way downstream, presenting a threat to boats that may be in the way and to certain types of installations on the banks. The spring floods when the ice goes out, often aggravated by heavy runoff from melting snow, are particularly damaging in the case of large north-flowing rivers like some of those in the Soviet Union. At Igarka on the Yenisei, for instance, the great lumber-loading facilities of the river port are said to need rebuilding after the big thaw each spring. Since the thaw on north-flowing rivers in the northern hemisphere begins upstream, the melt-water and floating ice spread over the still frozen river farther down, intensifying the threat to bridges, bank installations, and settlements on the flood plain. In the United States some of the worst floods on the Mississippi, the Missouri, the Ohio, and the Red River (of the North) have followed exceptionally rapid spring thaws.

Other types of problems are associated with the frozen ground in regions of long, cold winters. Water mains have to be buried several feet deep, and damages to home water lines keep the plumbers busy and prosperous. Surface mining may have to be interrupted for several months each season. *Frost heave,* the buckling of the ground surface as it freezes and thaws, presents a major challenge to highway engineers. Permafrost, permanently frozen ground, demands special techniques and added construction costs for airfields, roads, and building foundations. Maintenance costs are especially heavy on mountain roads in regions with frozen ground and heavy snowfall followed by spring floods.

Glaciers, formations of ice on the land which persist throughout the year, are of two general types: (a) *ice caps,* or *ice sheets,* which occupy certain broad upland areas and, near the poles, some lowland areas as well and (b) *valley glaciers,* which are streams of ice moving slowly down highland valleys in various latitudes. Somewhat intermediate between these two general types are *piedmont glaciers,* which occupy broad lowlands along the margin of a highland region in a topographic situation that permits two or more valley glaciers to coalesce or merge. The Malaspina and Bering glaciers along the coast of Alaska near the 60th parallel north latitude and the Frederikshaab Glacier on the west coast of Greenland are good examples of the piedmont type.[10]

Even though ice is a solid, it responds to pressure through changes in the crystalline grains of which it is composed and through an infinitesimal number of minor slippages along shearing planes. Another factor in glacial movement is the persistent thawing and recrystallizing of the ice, especially important in the lower, or wastage, portion of a valley glacier. Careful observation of the behavior of small ice sheets in various parts of the world indicates that a mass of firn or névé will begin to flow when it accumulates to a depth of 100-250 feet. The actual process of flow is almost impossible to observe because it is taking place beneath a considerable depth of overlaying ice. The movement of a valley glacier is greatest near the base of the center of it, while the very bottom and sides move much more slowly. The slower rate of flow here is due to the friction between the ice with its embedded rock material and the rock of the valley walls and valley bottom against which it is abrading.

Observations of the rate of flow of valley glaciers have been made in the Alps for at least two hundred years, often by the simple means of driving a straight line of visible stakes across the ice and then watching the straight line change to a parabolic curve as the center moves faster than the sides. Considering only the maximum surface flow near the center of the glacier, a rather typical rate of movement might

---

[10]Richard F. Flint, *Glacial Geology and the Pleistocene Epoch,* New York: John Wiley and Sons, Inc., 1947, pp. 13-14.

be about one or two feet per day, while in steeper portions of the valley this rate might be increased to ten feet or more per day. Exceptional instances have been observed in Greenland, in Alaska, in the Andes, and in the Himalayas of valley glacier movement of more than 100 feet per day.[11] Because of the greater rate of wastage at the lower end of glaciers in summer, it is quite difficult to study seasonal variations in rate of flow. A certain equilibrium ordinarily exists between the annual flow and the annual wastage of a valley glacier over the short run, but in a time span of hundreds or thousands of years the thickness and length of a valley glacier may vary enormously. The erosion of ice-carried boulders against the rock of the valley bottom and valley walls can eventually change the longitudinal and transverse profiles of the valley in striking fashion. An impressive amount of rock debris is left behind by a "retreating" glacier and the water flowing out from it.

The length of a valley glacier depends on its topographic situation and on the prevailing climatic conditions. Glaciers presently existing in the United States, even in Glacier National Park, are very small, although they were enormously larger in the geologic past. In higher or more humid mountains at the same latitude and in glacier regions nearer the poles, they are much longer. The Aletsch Glacier of Switzerland, longest in Europe, extends for about ten miles, while the Fedchenko Glacier in the high Pamirs of Soviet Central Asia is about 50 miles long. The Hubbard and Logan glaciers of Alaska approximate 75 miles in length and surely are among the longest in the world.

Glacial ice at the present time covers about ten per cent of the world's land area, but there is good evidence that, at its maximum extent in the Pleistocene period, ice covered approximately 31.5 percent of that same area.[12] There were still earlier "ice ages," it appears, in the Permo-Carboniferous period of earth history and in remote pre-Cambrian time, but the details of these earlier glacial periods are much less well-known that those of the Pleistocene which began one to two million years ago and continued, with drier, warmer interglacial stages of massive melting and deposition, almost to the present time.[13] Indeed, the ice caps on the Antarctic continent and Greenland, which account for about 85 percent and 11 percent respectively of the remaining ice, can quite properly be considered as what is left of the great Pleistocene ice sheets. In saying this, however, one needs to remember that

---

[11]Robert P. Sharp, *Glaciers,* Condon Lectures, Oregon State System of Higher Education, Eugene, Oregon, 1960, p. 37.

[12]Richard F. Flint, *Glacial and Pleistocene Geology,* New York: John Wiley and Sons, Inc., 1957, pp. 50-53.

[13]Clifford Embleton and Cuchlaine A. M. King, *Glacial and Periglacial Geomorphology,* New York: St. Martin's Press, 1968, p. 9 and ff.

they are undergoing constant renewal, and because of their continued *calving* into the oceans on their margin they were probably not much larger in the Pleistocene than they are at the present. The largest mass of floating ice presently attached to the land is the Ross Shelf Ice off Antarctica, which covers an area of about 200,000 square miles.

The great ice caps which covered so much of northern North America and northwestern Eurasia in the Pleistocene period, with ice depths of several thousand feet at their center, are now largely gone. But the evidence of their presence is everywhere: in the U-shaped mountain valleys, in the concave slope of numerous highlands, in the complex pattern of glacial lakes and swamps, in the great depths of bouldery moraine with their outwash trains of gravel and sand, and the vast sheets of dust-like loess which accumulated during the drier interglacial stages. The volume of ice still in existence on the land would, if suddenly melted, raise the water level of the present ocean area approximately 59 meters, or almost 200 feet. At the Pleistocene maximum the ice may have amounted to the equivalent of a column of water over the present oceans of 100 meters.[14] Considering the known fluctuations of climate and glacial ice, one cannot predict with any certainty that the ice will not come back or, for that matter, that the remaining ice will not all melt some time in the future and force hundreds of millions of people to move to higher ground.[15]

Economic Significance of Snow and Ice

The economic significance of snow and ice is evident in many spheres of human activity. Snow, by reflecting some of the sun's rays back into space, reinforces in some measure the cold weather which makes the snow possible in the first place. Since snow on the ground is an effective insulating layer, it may serve to protect tender plants (like fall-sown grain) and afford shelter for wildlife, while at the same time hiding surface subsistence for wildlife and open-range livestock. Rapidly-melting snow is responsible for some of the most damaging spring floods, but at the same time it contributes moisture to the topsoil that is almost indispensable in certain farming regions. Sleet is a well-known threat to trees and utility lines, and hail causes millions of dollars worth of damage to crops every year. The varied repercussions of heavy, drifting snow in the Great Plains of the United States have been well told in a professional article by Wesley Calef,[16] and

[14]Flint, *op. cit.*, pp. 54-55.

[15]H. W: Son Ahlmann, *Glacier Variations and Climatic Fluctuations,* New York: The American Geographical Society (Bowman Lecture Series), 1953.

[16]Wesley Calef, "The Winter of 1948-49 in the Great Plains," *Annals of the Association of American Geographers,* vol. 40, December, 1950, pp. 267-292.

FIGURE 4.2 Skiing on Mt. Rainier. Courtesy National Park Service. Snow comes early and lasts for many months on the middle slopes of Mt. Rainier in the state of Washington. This is one of the favorite winter sports areas in the Pacific Northwest.

the impact of a complex storm on the western part of the country has been portrayed in vivid detail by the novelist George R. Stewart.[17]

Water supply from melting snow and ice in the mountains is of particular importance to adjacent rural and urban settlements in many parts of the world. Cotton farming in such diverse areas as the San Joaquin valley of California, the Peruvian coastal desert, the Indus valley of West Pakistan, and the Amu Darya and Zeravshan valleys of Soviet Central Asia is utterly dependent upon irrigation from meltwater. In western United States, cities like Los Angeles, Denver, and Provo look to mountain reservoirs fed by melting snow for much of their municipal water supply.

Transportation is always affected by snow and ice conditions, sometimes favorably but more often unfavorably. In forested regions of high latitudes where there are few or no hard-surface roads, frozen river

[17]George R. Stewart, *Storm,* New York: Random House, Inc., 1941.

surfaces may serve as natural roadways for sledges or wheeled vehicles. Cross-country roads through the northern coniferous forest and tundra wilderness areas are often more trafficable in winter when the ground is frozen than in summer when the surface is unstable and the mud is deep. In mid-latitudes, however, snow and ice conditions present hazards to transportation by motor road, railroad, aircraft, or boat. Snow clearance from streets, highways, railways, and airport runways requires the prompt use of expensive heavy equipment, and even when this equipment is available there are inevitable delays in the movement of people and goods.[18] When canals, rivers, straits, salt-water seas, and ocean margins freeze over, even though submarines can still go under the ice and ice breakers can often force a way through on the surface, most water traffic comes to a halt. The shipping season in the Gulf of Finland, the Gulf of Bothnia, the St. Lawrence Seaway, and the upper Great Lakes is regularly shortened by ice. The search continues for equipment of a sort that will make regular shipping through the long-sought Northeast Passage and Northwest Passage technically possible and economically feasible.

Snow and ice are detested by many people to the point that areas with mild, sunny winters are showing magnetic attraction for population and industry in a mobile, competitive society. Nonetheless, they offer many other people recreation or associated employment. The pleasure and profit in the use or sale of shiny ice skates, skis, snowshoes, sleds, and toboggans is nothing new in communities with cold winters. What is newer, equally pleasurable it appears, and vastly more profitable per unit, is the cross-country snowmobile, which cost $600 to $1,000 or more. Its phenomenal popularity is sweeping Canada and northern sections of the United States. Winter sports in particularly attractive locales have bolstered the transportation industry, the hotel industry, and a host of industries making special clothing and equipment. The Winter Olympics regularly attract a large international clientele, and such winter sports centers as Stowe, Aspen, Sun Valley, St. Moritz, Chamonix, and Cortino d'Ampezzo are world-famous among the youthful-minded, energetic, and impressively affluent ski set.

[18]John F. Rooney, Jr., "The Urban Snow Hazard in the United States: An Appraisal of Disruption," *Geographical Review*, vol. 57, October, 1967, pp. 538-559.

Chapter 5

Rivers

The Origin of Rivers

A *river* is a major artery of a natural drainage system on the surface of the land. It ordinarily carries water, but the term river is also applied to the natural channel whether or not there is any surface flow present. It is used for both large and small streams, although frequently the smaller tributaries of a drainage system are called *creeks, brooks,* or (in some areas) *glens*. In the American Southwest, the Spanish term *arroyo* may be used for a stream which flows only after the occasional heavy shower.

Most of the earth's land surface is occupied by river systems, but about 16 percent has been calculated to be *areic,* that is, to have no surface flow of local origin, no permanent fresh-water lakes, and no regular renewal of ground water from local sources. In crossing such areic regions, rivers originating in more humid source areas lose volume and may disappear completely, leaving only an accumulation of undesirable salts on the surface. The areic regions of the earth include sizable portions of the mid-latitude and low-latitude deserts.

In addition to the 16 percent of the land surface that is areic, another 11 percent, according to calculations of DeMartonne and Aufèrre, is *endoreic*; this means that it drains only into lakes or marshes that are not themselves drained into the oceans. Endoreic areas would include the *drainage basins* of Great Salt Lake, Lake Chad, the Caspian and Aral Seas, and many lesser lakes, especially in the highlands of central Asia.[1]

[1]Emmanuel de Martonne and L. Auferre, "L'extension des regions privees d'ecoulement vers l'Ocean," *Annales de Geographie,* vol. 37, January, 1928, pp. 1-24.

Of all the precipitation which falls on the earth's land surface, an estimated 72 percent returns to the atmosphere through evaporation. The remaining 28 percent moves through river systems to the sea. The rivers are fed primarily by direct runoff from the land surface, but part of their water comes from seepage and springs. In certain areas, especially those that are or have been subjected to glaciation, the water which reaches the sea through river courses has been delayed for long periods, perhaps years or even centuries, by having been incorporated in an ice cap, a valley glacier, a lake, or a swamp. Many factors affect the relationship between precipitation and *runoff.* The initial evaporation loss to the atmosphere is influenced by temperature, wind, and humidity conditions. The fraction of the precipitation which soaks into the ground and makes little, if any, contribution to the eventual river flow is influenced by the porosity of the surface and subsurface, the degree of prior saturation of the surface and subsurface, and, of course, the slope.

The ratio of *river discharge* to precipitation may be as high as 50 percent or more in small river basins in humid mountains where there is much snow storage and rapid runoff from steep slopes. Most drainage basin discharge rates, however, are much lower. In plains regions of middle latitudes, there is little or no runoff from areas with less than ten inches of precipitation per year. In the subtropics and tropics, runoff may be nil or nearly so from areas of 20-40 inches of annual precipitation. The Mississippi River system includes tributaries from a wide variety of environments, as reflected in their discharge ratios. Those rivers entering the main stem of the Mississippi from the east have an average discharge ratio of 30 percent, while those entering from the drier western part of the system have average ratios of only five percent. Within the state of Oklahoma, where the mean annual precipitation ranges from about 50 inches in the southeast to 15-20 inches in the western panhandle, the surface runoff varies from more than ten inches in the southeastern counties to less than one inch in the semiarid west.

The presence of a lake, a swamp, or a series of lakes and swamps along the course of a river not only greatly retards the movement of water to the sea from the place where it falls as rain, but it also serves to make the river flow more uniform. This can be either an advantage or a disadvantage from the human use standpoint. It is a disadvantage when the upstream lakes and swamps provide large evaporation surfaces, letting water escape back into the air which is needed downstream or could be more efficiently stored for use downstream. Many writers have called attention to the enormous loss of water by evaporation from the Sudd swamp in the southwestern Sudan through which

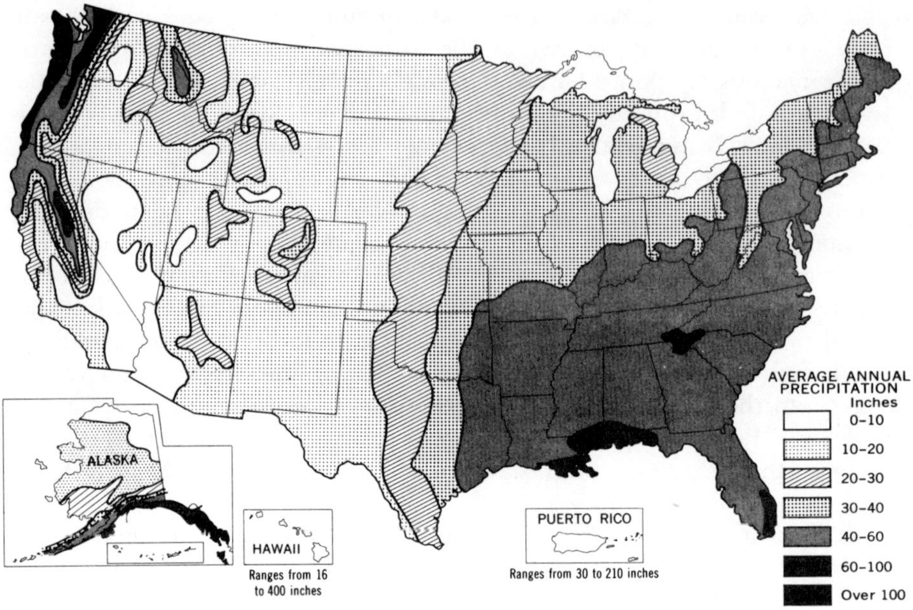

FIGURE 5.1 Average Annual Precipitation in the United States. In a given year the actual precipitation at a particular weather station in the humid eastern part of the country may vary from 50% to 150% of the station normal. In drier areas the year-to-year variability is much greater. From United States Water Resources Council, **The Nation's Water Resources,** p. 3-2-2.

the White Nile passes on its ways to join the Blue Nile and form the great river which nourishes thirty million Egyptian people. Waste of Nile River water is a matter of direct concern farther downstream. Russian engineers have noted that Lake Sevan in the Armenian Soviet Socialist Republic of the U.S.S.R. loses 838 millimeters (33 inches) of water each year from evaporation, 92 percent of all the water flowing into it. In order to assure that more than eight percent of the water of this high level lake (6,280 feet above sea level) is put to productive use, the engineers have designed a scheme of development for the Razdan (or Zanga) River which flows out of it that will lower the outlet somewhat and thereby reduce the evaporating surface.

An advantage of upstream lakes and swamps in a river basin, on the other hand, is that they may help to reduce flooding after periods of heavy rainfall, improve the dry season navigability of the lower course of the river, and help to sustain the water table at a desirable level in adjacent regions. The United States government has spent enormous sums of money to create artificial reservoirs along stream courses to achieve these very purposes and, in addition, to safeguard wildlife, improve water supply, and enlarge recreational opportunities.

In Chapter 4 the subject of ice and snow was related to the general problem of river flooding. It should now be noted that the contribution of melting snow and ice to the volume of flow of many rivers is essential to the established economy of downstream areas. The municipal water and hydroelectric power supply of numerous communities in and near the highlands of Scandinavia and the Alps is dependent upon meltwater. In the early part of this century Professor James E. Church of the University of Nevada worked out an ingenious and quite successful technique for calculating meltwater runoff by measuring the depth of the snow cover in various parts of the catchment basin and weighing the mass in each column, or boring, of snow sampled. At least eleven of U.S. western states now conduct systematic snow surveys in the late spring with a view to estimating the forthcoming summer water supply.

Seepages and *springs* are important sources of water for some rivers, perhaps especially for smaller ones. The emergence of water from a

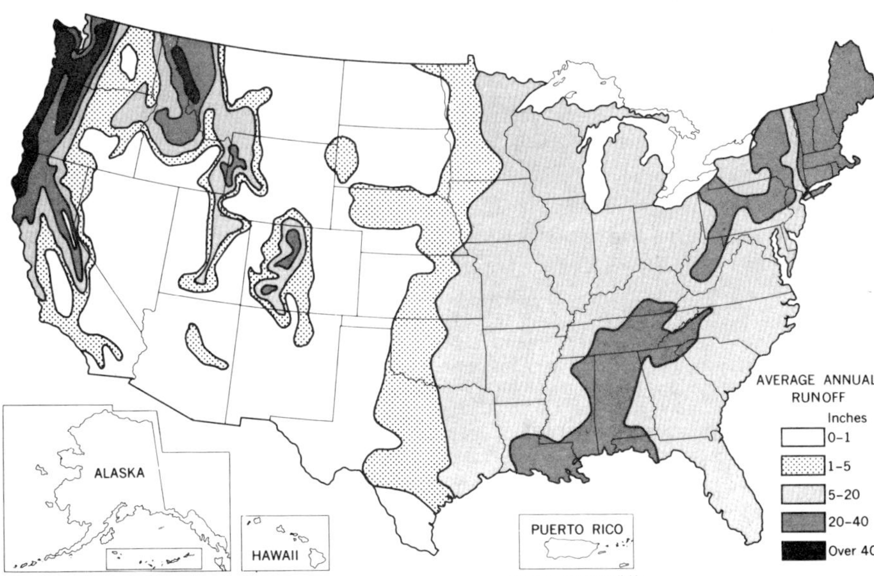

FIGURE 5.2 Average Annual Runoff in the United States. Surface runoff is many times greater in the humid eastern part of the country than in the West, where nearly all of the runoff originates in the mountains. From United States Water Resources Council, **The Nation's Water Resources,** p. 3-2-5.

saturated zone of the mantle rock or bedrock through which a river has eroded can sustain a flow in a small stream which would otherwise dry up shortly after each rain. Such seepages and springs are especially characteristic of humid regions where the water table is high, but they occur here and there where the nature and disposition of the rocks is favorable, even in semiarid regions and deserts. In situations where the underground water has been in contact with rocks high in lime, gypsum, sulphur, or salt, the emerging water may be distasteful, unpalatable for people and animals, or even totally unfit for agricultural use. More often, the contribution of seepages and springs to a surface water supply is a highly welcome and beneficial one.

Volume and Seasonality of Flow

The amount of runoff from a specific *watershed,* or drainage area of a river, varies from day to day, month to month, and year to year. In flash flood situations, following a heavy downpour of rain, it may vary significantly from hour to hour, or even from minute to minute. While the normal or average flow of a river is meaningful in unitary or comparative studies, it is often even more necessary to have data on general periodicity of flow and on the minimum and maximum flow which might reasonably be anticipated in a certain period of time. The runoff which will reach the main stem of a river system following a heavy rain or a period of rains depends not only on the amount and intensity of the rainfall but also on the prevailing temperatures and wind conditions, the slope, the dryness or wetness of the ground surface, the porosity of the surface, the nature of the vegetative cover, and other factors. The U.S. Army Corps of Engineers, in studying the flood control problem on a specific watershed in this country, makes a serious effort, using whatever data is available, to calculate the worst flood conditions that can reasonably be anticipated in a given period of time and refers to the level of "50-year floods," "100-year floods," and the like.

While the water in small streams high on a mountain slope is often quite pure and free of solid particles, a certain amount of sediment is characteristic of most river water. The quantity of sediment present is particularly high in *flash floods* in generally dry regions where there is a thin vegetative cover, but it is also very high in the surface runoff from basins in agricultural regions where a large fraction of the land surface is in intertilled crops. The increased volume and velocity of streams when they are in flood enormously increases their capacity to erode the *channel* and valley sides. It is during such flooding periods that the large boulders in a mountain stream shift their position and that farther downstream on the plain the erosion of the banks and shifting of the channel occurs, *meanders* are cut off, and occasionally a totally new route to the sea develops.

A river with a very low gradient may deposit sediment on the bottom and sides of its channel to the point that it creates *natural levees* for itself and flows several feet above the general level of the valley floor. Some of the worst floods on record on such rivers as the Mississippi and the Hwang Ho have occurred following a break in a natural levee. Unless the coastline of the continent is sinking in reference to sea level through slow tectonic movement, a river entering the ocean will extend the land by constructing a *delta* where it loses velocity and deposits its load. The Mississippi, the Mackenzie, the Yangtze, the Nile, and the Lena are examples of large rivers which have built extensive deltas. Similar deltas, but on a smaller scale, are often built by rivers where they enter lakes or artificial reservoirs.

The spatial arrangement of the various parts of a river system depends on such factors as the age of the river, the general slope of the drainage basin, the porosity of the mantle rock and underlying bedrock, the pattern of bedrock exposures within the basin, and the general elevation of the basin in relation to that of adjacent basins. Most rivers probably begin their lives as minor streams occupying incidental irregularities on an uneven surface somewhere near a coastline or a base level in an endoreic region (*i.e.*, a basin floor). As gullying develops, each stream lengthens headwards, competing for tribute from adjacent interfluves. Here and there, a tributary of one river, by having eroded its bed more deeply, is able to "capture" or divert to itself a tributary of another river. By enlarging its volume of flow, the river making the capture is enabled to deepen its channel and will proceed with headward erosion and additional *river captures*. The most successful rivers (if one may apply such an anthropomorphic term as "success" to rivers) have established their dominance over basins hundreds of thousands of square miles in extent and up to three thousand miles or more in length.

The drainage pattern of most river systems is *dendritic* or tree-like with a main stem, major branches of this stem, and a complex network of tributaries. In certain areas of ridge and valley topography, which have developed through differential erosion of tightly-folded and steeply-dipping sedimentary rocks of varying hardness, the drainage pattern becomes trellis-like in design. The *trellised drainage pattern*, as it appears in the Appalachian ridge and valley region of eastern United States or the Jura Mountains region of Switzerland and France, includes several nearly parallel major streams in the valleys with short tributaries entering them at right angles from the slopes of the intervening ridges.

As the main stem of a larger river system approaches *base level*, the lowest level to which it can erode with its present volume and velocity, it widens its channel and may, as an accompaniment to channel deposition, divide into several subchannels which form and reform with each successive flood. The Platte River in Nebraska is a textbook example of such a *braided stream*. Complex braided streams, sometimes called *distributaries*, are typical river forms on large deltas.

Most rivers begin small at their source and become progressively larger in volume of flow and width of channel as they approach their mouth. Exceptions to this pattern are apparent on the map in the case of rivers which rise in the mountains or in humid regions and move down slopes into dry plains where evaporation, *insoak*, and/or diversion for *irrigation* lessen their size and even lead to their total disappearance. In addition to a few large ones like the Nile and the Murray-Darling system of Australia, many of the rivers in endoreic regions are of this type. The Humboldt River of Nevada and the Zeravshan River which waters the ancient oasis city of Samarkand in the Uzbek S.S.R. are examples of rivers which lose themselves in *alkaline flats*.

The term "lost river" or "disappearing river" is most commonly applied to rivers flowing through soluble limestone country which find a solution channel beneath the surface and follow it for some distance, perhaps for several miles, before re-emerging on the surface. Such lost rivers are not unusual in regions of *karst topography* such as the Dinaric Alps of Yugoslavia or parts of Kentucky, Indiana, and Florida in this country. *Limestone caves* or grottoes with lost rivers in them are often visited by large numbers of tourists.

Many kinds of mechanical devices have been used to record river flows and flood-stage levels, from numbered gradations painted on a wall or a plank fastened upright in the water to complex recording instruments mounted in protected shelters. The U.S. Geological Survey operates more than 6,000 stream-gaging stations for recording volume of flow and 300 for measuring sediment load. The runoff record of United States rivers during the past several decades is available in various publications of the Geological Survey, primarily its voluminous *Water-Supply Papers* series. Mean daily, mean monthly, and mean annual discharge records are published, as well as other information concerning extremely high- and low-water stages.

Major rivers of the world can be compared as to length, area of drainage basin, and mean volume of flow in TABLE 5.1 which is based on various encyclopedic sources.[2]

[2]River length and area of drainage basins from *Encyclopaedia Britannica*, Chicago: Encyclopaedia Britannica, Inc., 1968 ed., Vol. 19, p. 354; mean discharge data from *Chambers's Encyclopaedia*, New Revised Edition, Oxford: Pergamon Press, 1967, vol. xi, p. 696.

TABLE 5.1

Drainage Basin Area, Length, and Mean Flow of Selected Rivers

River and Location	Drainage Basin (Sq. Mi.)	Length (Mi.)	Mean Flow (1,000's of Cu. Ft. Per Sec.)
Nile (Africa)	1,100,000	4,157	56 (at Cairo)
Amazon (South America)	2,722,000	3,915	3,300
Mississippi-Missouri (U.S.A.)	1,243,700	3,860	700
Ob-Irtysh (U.S.S.R.)	1,131,000	3,461	440
Yangtze Kiang (China)	756,498	3,434	800
Hwang Ho (China)	486,486	2,901	no data
Rio de la Plata system (South America)	1,680,000	2,796	970
Congo (Africa)	1,425,000	2,716	1,500
Lena (U.S.S.R.)	936,293	2,653	550
Mackenzie (Canada)	682,000	2,635	500
Niger (Africa)	580,000	2,600	500
Yenisei (U.S.S.R.)	1,003,474	2,566	615
Amur (U.S.S.R.)	758,000	2,552	386
Volga (U.S.S.R.)	533,818	2,293	287
St. Lawrence (Canada, U.S.A.)	291,900	1,900	400
Brahmaputra (Asia)	361,000	1,800	620
Indus (Asia)	372,000	1,800	no data
Danube (Europe)	315,444	1,777	no data
Zambezi (Africa)	513,500	1,700	no data
Murray (Australia)	414,253	1,609	14 (at Renmark)
Orinoco (South America)	350,000	1,281	440

It is quickly apparent from the tabular data that there is no close relationship between length of river, size of drainage basin, and annual flow. The Nile, longest river on earth, has only the sixth-largest drainage basin and is surpassed by numerous rivers in volume of water delivered to the sea. The Amazon is clearly the giant among world rivers in terms of the area it drains and the volume of water discharged. Early explorers of the east coast of South America found the Amazon dumping so much fresh water into the Atlantic that potable supplies could be taken from the ocean surface fifty miles from its mouth. Precise data on the volume of flow of most of the world's rivers is simply not available, and the figures which are published often need to be regarded as little more than crude estimates.

It is a rare river which maintains even an approximately uniform flow during the course of the year. The Deschutes River of Oregon, fed primarily by seepage from the porous lavas of the Columbia Plateau, and the Loup River system of Nebraska, supplied by springs and seepage from the porous sands of the Sand Hills region, are two of the better examples in this country. Even the St. Lawrence River, whose flow is regularized by the Great Lakes, shows marked variation in its volume of flow from season to season.

Quite apart from floods, which will be discussed later as a special phenomenon, most rivers portray a somewhat regular and predictable cycle of activity during the year. The cyclic changes in river flow relate primarily to variations in seasonal precipitation and in temperature, humidity, and wind conditions, which in turn affect evaporation and transpiration rates, insoak possibilities, ground freezing, snow and ice accumulation, and subsequent melting and runoff schedules. In certain rivers the seasonal flow at any one point is considerably affected by the activity of man upstream: by diversions for irrigation or other consumptive uses, by storage in reservoirs, or by releases of water from reservoirs.

Some European hydrologists have given close attention to seasonal flow cycles and have established definite types of *river regimes.* Maurice Pardé, for example, recognizes four basic categories of northern hemisphere river regimes:[3] (a) a *glacial regime* under which rivers fed by melting glaciers reach their maximum flow in July and August and their minimum (if any) flow during the coldest months of the winter; (b) a *rainy maritime regime* under which the low evaporation of winter in a situation of relatively even or winter-maximum precipitation brings the greatest runoff in the period from December to March and the least in August and September; (c) a *tropical rainy regime* under which the rivers rise to a maximum during and just following the time of maximum solar insolation and convectional rains and have their minimum flow at the end of the winter dry season, normally March and April; and (d) a *snowy plains regime,* characteristic of humid continental climates in middle latitudes where the spring snow melt, followed frequently by a period of intense local convection and thunderstorms, brings a runoff maximum in the March to June period and a minimum during the coldest months of winter. Some large rivers, like the Mississippi-Missouri, the Rhine, and the Danube have mixed regimes because the controlling factors differ significantly from one part of the river basin to another.

Floods are destructive, high-water stages of river flow. They are virtually unknown on some rivers, relatively frequent on others. Most natural floods in river valleys are the result of unusually heavy rainfall, unusually prolonged rainfall, or unusually rapid snow melt over a watershed or a sizable portion of a watershed. *Ice jams* on a river may

---

[3]Maurice Parde, *Fleuves et rivieres,* Paris: Librairie Armand Colin (Collection Armand Colin, No. 155, Section de Geographie), 3d ed., 1955, pp. 79-134.

be a contributing factor. Floods are characterized by an overflowing of the normal river channel and an invasion by water of the adjacent *flood plain,* with resulting damage to fields, fences, utility lines, roads, bridges, and buildings. They may destroy animal life and human life, even on a large scale. River floods are sometimes the result of the failure of an upstream dam built by man or a break in a temporary dam created by an avalanche, a mud slide, or a lava flow.[4]

The term flood is also applied to the invasion of the land by salty sea water, which occurs along coasts and the *tidal mouths* of certain rivers as a result of unusually high tides, strong onshore winds, or some combination of the two. *Dikes* or sea walls have been built to protect many shorelines and estuary margins against such natural disasters, but they do not always prove adequate.

While accurate data on the mean flow and normal cyclical variation of rivers is often inadequate or even non-existent, numerous flood records are available. On the Hwang Ho, the Yellow River, of North China, which has had disastrous floods in modern times, one flood record goes back to 2297 B.C. Flood records on the Nile have been kept for more than two thousand years and on the Danube for almost a thousand. The highest flood over the period of record on the Danube was in 1501 and on the Seine at Paris in 1658.[5] Hoyt and Langbein have compiled from a variety of official and unofficial sources a listing of some 2,000 floods which have occurred on 600 different waterways in the United States, starting with the one on the Mississippi River in the spring of 1543 which made it difficult for Hernando de Soto and his party to get across with their horses.[6] For those who read French, André Guilcher, a professor of hydrology at the Sorbonne, has compiled and shown graphically the mean behavior and flood records of a large number of rivers in various parts of the world.[7]

Some of the worst floods of the twentieth century have resulted from typhoon or hurricane winds or from *tidal waves* set in motion by underseas earthquakes, driving water onshore and engulfing coastal settlements. Japan, China, and India have suffered particularly heavy losses from floods of this type. The combination of gale winds and high tide on the North Sea in February, 1953, broke dikes and flooded extensive lowlands in the Netherlands, Belgium, and England, taking a death toll of more than 2,000. In the United States some of the most damaging

---

[4]Edgar E. Foster, *Rainfall and Runoff,* New York: The Macmillan Company, 1948, pp. 351-353.

[5]William G. Hoyt and Walter B. Langbein, *Floods,* Princeton, N. J.: Princeton University Press, 1955, pp. 5-6.

[6]*Ibid.,* pp. 331-431.

[7]Andre Guilcher, *Precis d'hydrologie,* marine et continentale, Paris: Masson, 1965, pp. 267-379.

river floods have been those on the Mississippi River in 1927 and 1936-37, the Missouri River system in 1951-52, and the southern New England and Pennsylvania rivers in 1936.

Special Characteristics of Selected Rivers

In order to illustrate some of the variety which exists among the major river systems of the earth, a paragraph-length sketch of several of them is introduced at this point. A book could easily be written about any one of them and, indeed, many books have been written about the geography, the history, and even the politics of individual rivers.

The Amazon, by far the largest of the world's rivers in volume of fresh water transported to the sea, rises in a chain of glacier-fed lakes in central Peru, within a hundred miles or so of the Pacific Ocean. Its drainage basin, including that of the Tocantins River which joins the Amazon's Río de Pará outlet on the south side of Marajó Island, is 2,722,000 square miles in extent, or about 90 percent as large as the United States excluding Alaska and Hawaii. While the principal tributaries rise on the north slope of the Brazilian Highlands, the southern slope of the Guiana Highlands, and the eastern slope of the Andes in Bolivia, Peru, Ecuador, and Colombia, the main stem of the Amazon courses through a nearly flat plain. The *tidal bore* reaches 400 miles inland, to the vicinity of Obidos. The head of navigation for ocean-going vessels, Iquitos, Peru, 2,300 miles inland from the mouth, is only 330 feet above sea level. Most of the basin, with an annual rainfall of 70 to 120 inches, supports a tropical rainforest, grading to savanna woodland toward the south. The Amazon basin is one of the world's great empty areas, with a population density averaging about one person per square mile. Many areas in the world have a water shortage. The Amazon carries out to sea, with little use along the way, a volume of water variously estimated at from four to seven times the discharge of the Mississippi, *and not far from one-fifth the discharge of all fresh river water from the entire land surface of the globe!*

The Congo, like the Amazon, is a tropical river draining a vast rainforest and savanna region. Because some of its major tributaries rise north of the equator and others south of the equator, the seasonal variation of flow which is so characteristic of these affluents tends to balance out in the main stem of the river. The drainage basin and volume of flow are about half that of the Amazon. Unlike the Amazon, which presents no hindrances to navigation for more than two thousand miles, the Congo has falls and rapids in its lower course. Above Matadi, the head of navigation, the river drops almost 900 feet in a series of falls and rapids which has been described as the greatest concentration of potential hydroelectric power anywhere in the world. Railroads have been built on both sides of this sector of the river to bypass the falls. Above Kinshasa and Brazzaville, the capital cities of the two Congo nations, the river is a broad and often braided one with an average gradient of less than five inches per mile, wandering amidst sandbars and with only about five to ten feet of usable depth. This portion of the basin is a saucer-like depression on the African plateau occupied by a shallow lake in recent geologic time. On the southern rim of the basin, the upper tributaries of the Congo also have falls and rapids, some of them developed for water power in the vicinity of the Katanga mining district. The population density of the Congo basin, while still a low 10-20 per square mile, is much greater than that of the Amazon basin.

The Nile River, considered to be the longest river on earth with a map-measured length of 4,157 miles, occupies a drainage basin covering about a tenth of the continent of Africa. Like the Amazon and the Congo, it rises in the tropics, but since its general course is south to north it flows out of the tropics and across the largest desert on earth before spreading out on its delta below Cairo and entering the Mediterranean Sea. Of the three principal tributaries, the largest is the Blue Nile, which drains the rainy western highlands of Ethiopia and contributes about four-sevenths of the Nile water. The White Nile tributary, which contributes about two-sevenths of the flow, drains a substantially larger area than the Blue Nile, but much of the potential runoff is lost by evaporation from Lake Victoria, Lake Albert, and the great El Sudd swamp. The Atbara tributary which drains part of northern Ethiopia contributes the final one-seventh of the total flow. Below the Atbara the Nile continues for almost half of its total length, over 1,600 miles, without any further tributary of consequence. There are zones of *rapids*, or *cataracts,* intermittently along the Nile, but long stretches of it are navigable by small vessels. The gradient of the river is slight enough that silt has been deposited over a broad flood plain throughout the middle and lower course. Systematic diversions of water from the Nile have supported an irrigation culture since ancient times, and today more than thirty million Egyptians are dependent on it for their food supply. The Aswan storage dam, built by British engineers early in the twentieth century, is being dwarfed by the Aswan high dam, under construction with Russian financial and technical assistance since 1960. When completed, the high dam will make possible a 20 percent expansion of the irrigated acreage. An agreement has been reached between the United Arab Republic and the Sudan for sharing the Nile water on which both nations depend.

The Hwang Ho, or Yellow River, of northern China, while almost as long as the Yangtze of central China (2,900 miles) carries much less water and much less traffic. It rises in a rather arid but ill-drained upland to the northeast of Tibet and follows a sinuous course far to the north, then east and south again, where it is joined by its principal tributary, the Wei-Ho, before breaking out of the upland through Tungkwan Gorge onto the North China Plain. In its final 500 miles or so it has a very slight gradient and has deposited on its bed until it flows 15 to 30 feet above the general level of the plain and is held within its banks only by natural levees. The north-to-south sector of the river, between Hokow and Tungkwan, occupies a very narrow valley in contrast to the broader, more mature valley both above and below this sector, suggesting a possible *river capture* sometime in the geologic past. Upstream from the Tungkwan Gorge the Hwang Ho flows for long distances through a semi-arid plateau of wind-blown dust, or loess, whose yellow color gives the river its alternate name. Another name sometimes given the Hwang Ho is "China's Sorrow," for on the North China Plain it has more than once broken through its natural levees and flooded wildly, destroying vast areas of grain and vegetable crops and taking thousands of lives. In 1852, in one of these floods, it shifted its whole lower course from an outlet south of the Shantung uplands to a new one on the other side of the uplands 250 miles to the north. The government of China is presently working on some ambitious river development plans, including two large hydroelectric projects, one near Lanchow and one in the Tungkwan Gorge area.

The Yenisei, a north-flowing river of Siberia, is the largest river of the U.S.S.R. in volume of water discharged, but it is surpassed in length by two other Siberian rivers, the Ob-Irtysh and the Lena. It rises, like the Hwang Ho, on the dry plateau of central Asia, breaks northward through a gorge out of the remote Tuvinian Basin into the industrially important Minusinsk Basin, then cuts through a granite chasm across the western Sayan Mountains to a broader plain at the city of Krasnoyarsk, where it is crossed by the Trans-Siberian Railroad. Recent engineering work on the upper course of the river has made it somewhat navigable all the way to Kyzyl in the Tuvinian Basin. Below Krasnoyarsk, however, river boats and barges ply with regularity during the summer season. Since there are no paralleling roads or railroads, the river is the only avenue of north-south movement apart from winter snow trails. In this middle and lower sector, the Yenisei marks the eastern edge of the swampy Ob-Irtysh basin, with the western margin of the Central Siberian Highlands rising prominently above the right bank. The drainage pattern here is curiously asymmetrical, with practically all of the tributaries entering from the adjacent highlands on the east. One of these tributaries, the Angara, drains Lake Baikal, and along its course is the giant Bratsk hydroelectric plant. Bratsk is exceeded in size among the hydro plants of the world only by another Yenisei plant (4,000,000 kw. capacity) nearing completion in the narrows above Krasnoyarsk. Logs and lumber are the principal items of downstream freight, with Igarka especially prominent as a center of shipment. Below the river port of Dudinka, from where a short rail line connects with the mining center of Norilsk, the river has been drowned by crustal subsidence and becomes a broad arm of the Kara Sea. Winter ice begins to form on the lower Yenisei in early October, and when the spring thaw finally sets in along the upper reaches of the river in May it causes appalling floods in the still-frozen lower parts of the great valley.

The Volga, largest and longest (2,293 miles) of the rivers of Europe, has no source in mountain glaciers or snowfields. It rises in the morainic Valdai Hills at an elevation of about 750 feet, then makes its way slowly eastward to the great bend at Kazan and from there southward to the Caspian Sea, with an average rate of descent of only four inches per mile. In its upper reaches it flows through a mixed forest region, only partly cleared for farming; from Kuybyshev to Volgograd it passes through prairie country, mainly sown to wheat; and below Volgograd it enters Europe's only real desert as a braided stream sprawled in a broad floodplain, to terminate in a huge delta at the north end of the Caspian Sea, more than 90 feet below sea level. In its natural state the Volga River portrayed the typical features of Pardé's snowy plains regime. The runoff at Kuybyshev at the height of the snow-melt season in May was about ten times the flow in the winter months when the surface of the river was frozen. At present the discharge pattern is a regulated one, for on both the Volga and its major tributary, the Kama, the Soviet government has built a series of hydroelectric power dams and multi-purpose reservoirs which retard the spring runoff and release it as desired. Upstream from Volgograd the main stem of the river all the way to its source is now little more than a series of man-made lakes. The Volga has always been a navigable river over most of its length, but the system of lakes and locks has greatly increased its navigability. Large canals link the Volga with the Don River and hence the Black Sea, with the Baltic Sea via Leningrad, and with Moscow. Diversion of Volga water for irrigation and the evaporation from the great reservoirs have caused concern about the falling level of the Caspian Sea. The Volga River has many large cities along its course, and more than fifty million people live within the drainage basin.

The Rhine River, largest of western Europe, is a small river in comparison with those previously described. It is slightly over 800 miles in length, less than a fourth that of the Nile or the Amazon, and scarcely more than a third the length of the Volga. Its volume of flow is only five to ten percent of that of the Congo. Even though there are no large regulating reservoirs along its course as there are on the Volga (the largest is glacial Lake Constance, called in German the Bodensee) the discharge of the Rhine is fairly uniform due to the contrasting regimes of its upper and lower tributaries. The upper Rhine above Basel has a mixed glacial and snowy plains regime with a May-June-July maximum, while the middle and lower tributaries (Neckar, Main, Moselle, and others), although contributing some melt-water, have a predominantly maritime regime with maximum discharge in winter. The Aare River, fed by melting Alpine snow and ice, more than doubles the flow of the Rhine at the confluence of the two streams between Lake Constance and Basel. Several of the upper tributaries, along which there are a number of hydroelectric power sites, have their origin in permanent snowfields and cirque lakes between 7,000 and 8,000 feet above sea level. Another series of hydroelectric plants, larger but still small in size compared with the giants on the Volga and the Yenisei, is located at and near Schaffhausen along the northern margin of Switzerland. Below Basel the Rhine has been extensively improved for navigation, and the volume of freight traffic on the German and Dutch sectors is among the highest, if not the highest, of all the rivers of the world. In the Netherlands the Rhine River water reaches the sea through a complex and carefully regulated network of natural and artificial *distributaries*. The Rhine valley has many historic cities, and the river basin is one of the most densely populated river basins in the world.

The Mississippi River is the largest river of North America, with a discharge estimated at eight times that of the Rhine. With its Missouri affluent, it is surpassed in length only by the Nile and the Amazon, and among rivers which rise outside of the tropics it is exceeded in volume of flow only by the Yangtze. The nomenclature of the Mississippi system is somewhat confusing in that the tributary originating in the glacial lakes of Minnesota, which gives its name to the main stem of the stream, is neither the longest of the major affluents (the Missouri) nor the one which contributes the most water (the Ohio). Altogether the Mississippi system drains much of the interior of the United States from the Rocky Mountains to the Appalachians. As noted earlier, the runoff from the eastern part of the basin is very much greater than that from the more arid western segment, where some of the affluents see much water diverted for irrigation

(Platte, Arkansas) and others become little more than intermittent streams during the late summer, autumn, and winter. The general regime of the Mississippi is the snowy plains type, with a spring-early summer discharge maximum, merging the snow-melt season with the time of maximum thunderstorm convection. Much of the Mississippi basin is crop land, and the volume of silt transported by some of the tributaries, especially the Missouri, is enormous. Since the Mississippi system carried a much larger volume of water during certain intervals of the Pleistocene period, the main stem and the principal affluents flow in broad valleys with wide flood plains and extensive *terraces.* For the last 1,000 miles or so of its course the Mississippi is flowing across a vast, elongated delta which is still extending into the Gulf of Mexico in spite of crustal subsidence. The river has been improved for navigation by dredging, channel straightening, and lock construction (Keokuk, Iowa), but the volume of freight carried is much less than on the Rhine. A large part of the Ohio-Tennessee system, however, has become quite intensively used, as there is much heavy industry in this portion of the basin. Damaging floods have occurred on various parts of the Mississippi system, and for the past forty years the U.S. Army Corps of Engineers and other government agencies have been engaged in a vast flood control operation which is still not completed.

The St. Lawrence River, largest of the North American rivers to enter the Atlantic Ocean, drains the surface of the five Great Lakes and the numerous short rivers which enter them. Farther downstream, below Lake Ontario, the St. Lawrence receives a rather sizable contribution from the Ottawa, Richelieu, Saint-Maurice, and Saguenay tributaries. For hundreds of miles below Quebec city the river is brackish and tidal, a broad estuary with banks so far apart that sixteenth-century explorers entering it felt sure they were on their way to China. The obstacle which forms the natural head of navigation just above Montreal was dubbed the Lachine Rapids. Short portages and long reaches of lake with interconnecting rivers made the Great Lakes-St. Lawrence route a natural entrant to the continent. It provided a base for French colonial development and eventually became a major focus of French and British rivalry on the continent. Because the Great Lakes serve to regulate the flow, the St. Lawrence varies much less between maximum and minimum stages than the Mississippi and its tributaries, on the order of only two to one. The four upper lakes are within thirty feet of being on the same level, but the twenty-foot difference between Lake Superior and the Michigan-Huron surface requires a large lock at Sault Ste. Marie to permit the iron ore boats and other freighters to pass through. The drop from Lake Erie to Lake Ontario along the

Niagara River (the stream becomes the St. Lawrence only below Lake Ontario) is about 325 feet, half of it in the sheer drop of Niagara Falls, a world-famed scenic attraction. The Welland Canal, built as a bypass around Niagara Falls, was first opened in 1829 and has subsequently been deepened and enlarged at least twice. One of the most spectacular engineering achievements on the North American continent, the Great Lakes-St. Lawrence Seaway project, was completed in 1959, providing a minimum channel of 27-foot depth all the way to Duluth and Chicago. Hydroelectric power is of major importance along the New York-Ontario sector of the Seaway and on the Saint-Maurice and Saguenay tributaries. One of the greatest industrial concentrations in the world lies on and near the shores of the Great Lakes and along the St. Lawrence River.

The Colorado River, which drains a thinly populated, Texas-sized area in southwestern United States (and should not be confused with the smaller Colorado River in Texas), debouches into the Gulf of California after an 80-mile course across the Mexican section of its delta. The Colorado bears a certain resemblance to such rivers as the Nile, the Tigris-Euphrates, the Indus, and the Amu Darya in that it rises in a humid mountain region, with ample rain and heavy snowfall, and then over a large part of its course flows through a desert region where it is losing rather than gaining water. The principal sources of the Colorado are high on the Wind River Mountains of Wyoming, the slopes of the Uinta Mountains and Wasatch Plateau of Utah, and the Middle Rockies of the state of Colorado. For more than a thousand miles, almost two-thirds of its entire course, the Colorado River flows in a deep gorge with no flood plain at all and with high canyon walls rising to a barren plateau surface 6,000 to 9,000 feet above sea level. While some of the upper tributary canyons are highly spectacular, such as the Gunnison in Colorado and the Green River gorge through the eastern Uintas, the most impressive of all these is the Grand Canyon in northern Arizona, where over a distance of 200 miles the river has cut a ragged, colorful gash more than a mile deep, exposing in the canyon wall rocks which represent almost two billion years of earth history. The natural flow in the lower Colorado River is said to vary from 3,000 cubic feet per second to more than 200,000 cubic feet. Upper portions of the river basin support a number of agricultural settlements, but the more intensive land use in the basin is below the canyon section in southern California (Imperial Valley), southern Arizona (Gila-Salt Valley), and adjacent sections of Mexico. Large scale hydroelectric power development on the Colorado River began with completion of the Boulder Dam in 1936. Lake Mead, behind the 726 foot high dam, has a length of over a hundred miles and a capacity of more than 30,000,000 acre feet. Other dams, both above and below Boulder Dam, store water for irrigation, power production, and municipal use. One of the unusual aspects of Colorado River development is the large scale diversion of water out of the basin for use elsewhere. Water moves from the reservoir above Parker Dam to the Los Angeles and San Diego area, and from the upper Colorado by tunnel through the continental divide to the Big Thompson Canyon for use in communities on the eastern slope of the Rockies. The existence of the Salton Sea, 240 feet below sea level in a structural depression of southern California, is the result of a broken river dike in 1905 which required two years and $2,000,000 to repair. For those who like to study political problems of water allocation and use, the Colorado River is a gem.

Economic Significance of Rivers

Rivers are of importance in human geography for a wide variety of reasons. They have afforded man marvelous opportunities and in other situations have presented him with formidable obstacles. Some rivers have been harnessed rather completely to serve the present wishes of man; others flow as wildly and freely as they did before he came on the scene. The economic significance of rivers to man will be described here very briefly under eight general headings: (a) navigation and routeways, (b) irrigation, (c) water power, (d) domestic and industrial water supply, (e) floods, (f) fishing and recreation, (g) pollution, and (h) boundary functions.

Rivers have served as routeways for the movement of people and goods since long before there were any historical records. The proof is in the exotic character of many of the objects excavated from ancient campsites along and near navigable streams. In earlier times, when little of the world's forest land had been cleared and overland roads were, at best, rough trails, rivers offered the easiest and safest avenues for migration and trade. In later times, towns developed along river courses, especially at river confluences and near the mouths of rivers where coastal traders could make easy contact with riverine peoples. *Portages* between navigable rivers became familiar routeways, and control of them often had real strategic significance. In modern times many of the natural waterways have been deepened, marked with buoys to improve safety, and provided with port facilities for loading and unloading, warehousing, and trans-shipment. *Canals* have been built in many places to interconnect river routes and to improve the access of river routes to the sea, just as the Suez, Panama, Kiel, and other canals have been built to shorten sea routes.

Inland waterways serve some regions of the world that have no alternate means of moving bulk goods. Coal, iron ore, logs, lumber, and other heavy building materials figure prominently in the traffic, along with petroleum and grain. The greatest concentration of inland waterways, perhaps surprisingly, is in areas already provided with alternate overland and air facilities, especially in Europe and the eastern half of the United States. The Rhine, Elbe, Danube, Dnieper, and Volga rivers all have a heavy long distance traffic, as do the Mississippi-Ohio-Illinois system in the United States and the St. Lawrence shared by the United States and Canada. In southeast Asia, the Yangtze, Mekong, and Irrawaddy rivers are heavily used as are the Parana-Paraguay and Magdalena in South America. The drowned mouths of some small or otherwise unnavigable rivers, often extended inland by *dredging* or canal construction, have seen the growth of great entrepôts. Among American examples are: the Columbia (Portland), the Sacramento (San Francisco-Oakland), the San Jacinto (Houston), the Patapsco (Baltimore), the Delaware (Philadelphia), the Hudson (New York), and the St. Lawrence (Montreal). In Great Britain and continental Europe there are parallel developments on the Thames (London), the Clyde (Glasgow), the Mersey (Liverpool-Manchester), the Tyne (Newcastle), the Scheldt (Antwerp), the Weser (Bremen), and other rivers. In Asia, Tientsin, principal port of North China, is more than 40 miles up the Pai-ho, and Calcutta in India is 60 miles inland from the mouth of the Hooghly, one of the distributaries of the Ganges. In all parts of the world, access to water navigation is one of the magnets which attracts people and encourages urbanization. Of the 60 or so cities in the United States with a population of 200,000 or more, nearly 80 percent are located on the nation's waterways.[8]

Earliest use of rivers for *irrigation,* like the beginning of river transportation, is lost in antiquity. Water engineering appears to have been a well-developed art in the Tigris-Euphrates and Nile valleys at least 5,000 years ago.[9] The need to cooperate in the management and utilization of water is believed to have been a powerful incentive toward the development of organized societies larger than the tribe or clan. Irrigation has been used in the cultivation of rice in China and India at least since the second or third millennium before Christ. With the increased availability of machinery for leveling and ditching land, for pumping, piping, and sprinkling water, and for building dams and reservoirs, modern man has greatly extended the area under irrigation.

[8]Clarence F. Jones and Gordon G. Darkenwald, *Economic Geography,* New York: The Macmillan Company, 3d ed., 1965, pp. 734-743.

[9]William H. McNeill, *A World History,* New York: Oxford University Press, Inc., 1967, pp. 11-29.

The rapid growth of population since the Industrial Revolution, especially of urban population without land on which to grow their own food, has been a spur to commercialized farming. An expanding market for cotton, rice and other grains, cane and beet sugar, specialized fruit and vegetable crops, and livestock fodder has pushed intensive agriculture into selected spots in the world's dry lands where surface water or ground water can be made available for irrigation. By far the largest fraction of irrigation water is secured through the storage and diversion of river water. The Ganges and Indus river systems were developed for irrigation on a large scale during the period of British rule in India. The ancient irrigation systems of China, Egypt, Iraq, and Soviet Central Asia have been modernized and, in many cases, extended. Numerous small rivers in southern Europe now support intensive irrigation operations. In the United States such western valleys as the Arkansas and Platte, the Snake and San Joaquin, the lower Colorado and the Rio Grande present similar scenes of intensive irrigation activity.

Water power is another major economic value of rivers. The undershot and overshot water wheel are ancient devices for grinding grain and doing other useful chores. The Industrial Revolution brought many new uses for the turning shaft, and the millpond with its water wheel became a familiar sight along numerous small rivers in Great Britain, western Europe, and New England. With the development of *hydroelectric power* in the late nineteenth century, however, and the associated construction of long distance transmission lines, water power really came into its own. Where the water supply is reliable, the gradient in the valley is sufficient, the rock structure is suitable for dam and reservoir construction, and there is a market for power at hand or within a reasonable transmission distance the appropriate conditions for water power development are present. Only a few countries, Switzerland, Norway, and New Zealand among them, generate most of their electricity with water power. All of the major industrial countries, such as the United States, the U.S.S.R., Japan, the United Kingdom, West Germany, and France, have important hydroelectric power developments, but in none of these countries does the output of hydro power equal that generated from combustible fuels.

Many of the newer, underdeveloped countries of the world have little or no output of hydroelectric power because of high initial construction costs, a lack of suitable power sites, or limited markets. In some of them the market for electricity is still very small and can best be met by modest-sized municipal plants burning imported fuel. Overall, then, there is little correlation between the world maps of potential and developed hydroelectric power. The countries with a large amount of developed hydro power are those with a large power market, while the vast potential resources in areas like the Congo basin and tropical West Africa, southeast Asia, Central America, and the Guiana Highlands of South America lie almost untouched.

Rivers and lakes, the latter often artificial ones, are the foremost source of domestic and industrial water supply. Per capita water for domestic and industrial use may be as little as a gallon or two a day in a country with a primitive subsistence or agricultural society and a hundred times that much in a highly industrialized urban society. Among the three-fourths or so of the United States population dependent upon public water supplies in 1960, per capita use was approximately 150 gallons per day. In addition to the amount used for drinking, cooking, and personal hygiene, this consumption represented extensive employment for laundry purposes, swimming pools, air conditioning, lawn sprinkling, car washing, and other family, community, and industrial uses. Some industries such as petroleum refining, iron and steel production, paper manufacturing, brewing and distilling, and certain branches of the textile industry are especially heavy users of water.[10] Most industrial uses of water, including the generation of hydroelectricity, do not consume water in the sense that irrigation consumes it. In a densely populated industrial valley the same water may be withdrawn from a river, used, and returned to the river several times. Making domestic and industrial water available in suitable quantities and with essential quality standards has become an expensive public responsibility in many communities. It is not unusual, however, for industrial firms to assume responsibility for their own water supply. *Per capita water consumption* is increasing in many other countries as well as in the United States, and there is growing concern about how to keep supply abreast with demands. *Recycling sewage water* and *desalinizing sea water* are two approaches now in limited use.[11]

Fishing in rivers is for some fishermen primarily a source of subsistence food, for others a type of employment, and for still others a form of recreation. It is an ancient activity of man which in some parts of the world continues to be an important means of livelihood. Such is the case, for instance, with many people who live in the tropical rainforest or in high latitude forest and tundra environments where grains are difficult or impossible to grow and most domestic animals do not thrive. Even in areas where food crops and domestic animals do thrive, fresh water fish may be an important supplementary source of high protein food. While 84 percent of the total world fish catch

---

[10]Peveril Meigs, "Water Problems in the United States," *Geographical Review*, vol. 42, July, 1952, pp. 346-366.

[11]Gilbert F. White, "Industrial Water Use: A Review," *Geographical Review*, vol. 50, July, 1960, pp. 412-430.

by weight is taken from the oceans and bordering seas, the remaining 16 percent, which represents the freshwater catch, is of enormous importance in parts of Africa and Asia.[12] Inland fisheries, while less essential to the regional food supply, are of significant commercial value in many parts of Europe and the Soviet Union. In Anglo-America the Great Lakes yield about half the commercial supply of fresh-water fish, with the other half taken from smaller lakes and streams. Numerous state fish hatcheries in the United States supply fingerlings to restock mountain trout streams, farm ponds, and larger rivers and lakes. The recreational value of rivers and other inland waters for fishing, boating, swimming, water skiing, and other sports is impossible to measure in monetary terms, but it is obviously very great.

While rivers have many positive economic values for man, in navigation, irrigation, hydroelectric power, water supply, fishing, and recreation, they may have negative values as well. A river out of control is a serious natural hazard, like lightning, tornadoes, hailstorms, hurricanes, earthquakes, volcanoes, locusts, or Dutch elm disease. A river in flood is a threat to life and property; the cost in physical damage alone from a major one may amount to several tens of millions or even hundreds of millions of dollars. As indicated earlier in the chapter in the discussion of flooding as a physical aspect of rivers, records of floods have been kept since ancient times. The true extent of loss in a flood is impossible to calculate. Modern man's urge to know what he is getting for his money—in other words, to calculate cost-benefit ratios from investment of funds—has led him to at least attempt estimates. Flood experts in the United States Weather Bureau have estimated flood losses in the United States (exclusive of Hawaii and Alaska where data was not available) at a little more than $6,500,000,000 during the period from 1925 to 1961, and they have calculated the share of these losses occurring in each of the major river systems of the country.[13] In the Missouri River basin alone, the estimated flood loss during the 36-year period under study was $1,797,000,000.

Human response to the occurrence of floods may be one of dumb resignation or it may be a more positive one. Positive responses require foresight and usually cost money—much money. Some experts have recommended a public system of flood insurance, and others have urged strict floodplain zoning to be enforced by public authorities. In the United States the principal response to the flood threat has been in the form of efforts to modify land use, as through reforestation, to reduce runoff by allocating funds to construct terraces on sloping crop

---

[12]Jones and Darkenwald, *op. cit.*, p. 33.

[13]Guy-Harold Smith, ed., *Conservation of Natural Resources*, New York: John Wiley and Sons, Inc., 3d ed., 1965, p. 298.

land, to build farm ponds and check-dams on minor tributaries and, under the federal Flood Control Act of 1936, to build large catchment reservoirs on major rivers. The multi-purpose concept now attributes benefits from the large dams not only to flood control but also to water power, water supply, and public recreation.

While floods are an ancient hazard to man and his works, *water pollution,* at least on a large scale, is a quite recent one. Pollution is a function of increased numbers of people, of new forms of economic activity, and to a disturbing extent is a result of human ignorance, indifference, carelessness, and misplaced priorities. When the population density of the world was much less and in all countries largely rural, the problem of pollution was a minor one and such a dispersed one that it attracted little attention. The big problems of man were war, famine, and disease, and, of these, disease was the least understood. Microbiology, which has contributed so much to an understanding of water pollution and its effect on health, is essentially a new science, a science of the past century. It is now known that pollution from human, animal, and industrial wastes has caused incalculable damage to rivers, lakes, and shorelines, and in places even to ground water and ocean surfaces. The proper disposal of wastes is a serious local, state, national, and international problem.

Approximately a hundred years ago it began to be obvious to public authorities that some river water in the United States was not fit to drink. The first filter plant in the country was built in Poughkeepsie, New York, on the Hudson River in 1875. By 1900 a number of cities had installed individual water works, and about three percent of the population was drinking purified water. More communities added treatment plants, and by now most Americans have access to reasonably pure water, but the pollution problem is still far from solved. Shortly after World War II, the U.S. Public Health Service reviewed pollution conditions in eleven major river valleys in various parts of the country. These field surveys covered 4,409 municipalities and 3,413 industrial plants. Of the cities studied, 43 percent, and of the factories, 57 percent were discharging raw untreated waste into nearby water. Some 800 of the treatment works in operation were found to be overloaded or otherwise inadequate.[14] The federal Water Pollution Control Act of 1948 provided loan funds to municipalities to enable them to construct necessary sewage and waste treatment facilities. The federal Water Quality Act of 1965 established national standards for water supply and pollution prevention in the form of guidelines to the states, but

[14]*A Water Policy for the American People*: The Report of the President's Water Resources Policy Commission, vol. I, Washington: Government Printing Office, 1950, pp. 185-195.

full implementation and the enforcement of penalties present formidable challenges. President Nixon, in his State of the Union Message in January, 1970, identified the reduction of water and air pollution as one of the major challenges confronting American society.

It seems appropriate to conclude this discussion of the human geography of rivers with some mention of their political significance. Because of their linear character on maps (often much more apparent on maps than on the ground) many rivers have been selected to serve as boundary lines between political units. Studies in political geography, political history, international relations, and international law are full of examples of the strategic value, real or presumed, of particular river lines. Modern France has always placed a high priority on having a frontier on the Rhine. Delicate diplomatic negotiations have been called for in recent years to partition the waters of the Nile, the Indus, and the Rio Grande. In Europe and Africa there are many instances of rivers serving a boundary function, and a large number of interstate boundaries in the United States are established on rivers.[15]

Without question, rivers in the past, even as recently as World War II, have served usefully as obstacles to the movement of military forces. Deep, canyon-type valleys are difficult to cross and may effectively separate different types of people. Rivers of low gradient in broad valleys, however, have generally furnished avenues of movement and have seen the development of similar cultures and interests on both sides. Experience suggests that there is little merit in fixing political boundaries along shallow streams with braided channels, shifting *thalwegs,*[16] and wide-swinging meanders. The El Chamizal problem in the vicinity of El Paso was a sticky one in U.S.-Mexican relations for several decades. Interstate compacts and international agreements concerning river navigation, water diversion, power development, and pollution control are already multitudinous. It is a safe guess that there will be many more of them in the future.

[15]Stephen B. Jones, *Boundary-Making: A Handbook for Statesmen, Treaty Editors, and Boundary Commissioners,* New York: Columbia University Press, 1945.

[16]The thalweg of a river is the line which marks the deepest channel.

Chapter 6

Lakes and Swamps

Origin, Types, and Distribution of Lakes

In contrast to a river, an essential characteristic of which is its gradient or slope on land, a *lake* is a relatively still body of water which occupies a depression on the land surface. Implicit in the concept of a lake is the notion that it is a body of water of substantial size, in contrast to a *pond* or a *pool*, either of which is smaller. The term *sea* is applied to certain very large lakes, like the Caspian and Aral, but to avoid thinking there is logic here, one should remember the Sea of Galilee which is only a few square miles in extent. Logical consistency in the nomenclature of water bodies is clearly too much to hope for, and, as was noted in Chapter 2, popular usage is too ingrained for the U.S. Board on Geographic Names or its equivalent in other countries to have much success in imposing name changes. So, even while reading definitions or perhaps attempting to create them, one must learn to live with such place-name inconsistencies as the Salton Sea and the Sea of Galilee, or Lake Pontchartrain and Sabine Lake along the Gulf of Mexico. Even though, as was admitted earlier in this book, one cannot be sure how many oceans there are, the state of Minnesota boasts with confidence and pride that it is the "Land of Ten Thousand Lakes."

Lakes are not only of various sizes, shapes, depths, colors, mineral characteristics, organic contents, and degrees of permanence; they also have many different types of origins. Some lakes are merely wider places in a river where there is some reduction of gradient, like Lake Pepin on the Mississippi below Red Wing, Minnesota. Others occupy meander scars where a river has abandoned a segment of a former channel and created a shallow dam by sedimentation along the side of its new channel. Lake Beulah, Lake Lee, and Lake Washington on the east side of the Mississippi River in the state of Mississippi are examples. Reelfoot Lake on the Mississippi River floodplain in northwestern Tennessee occupies a shallow basin on the floodplain caused by a sudden subsidence of the surface at the time of the New Madrid earthquakes in 1811-12.

The essential common feature of all lakes seems to be that they have a basin of some sort, large or small, deep or shallow, which retards, interrupts, or prevents the flow of surface water to the ocean. Lakes are temporary phenomena on the earth's surface and, as one writer so vividly put it, "rivers are their mortal enemies." Most lakes can be expected to disappear in a very short time, at least in the geologic sense, either through the eroding away of their outlets or the filling of their basins by sedimentation. While recognizing that many lake basins have more than a single cause, it is rather customary to classify lakes according to the predominant way in which their basins were formed. Six general types will be recognized here: (a) lakes formed by tectonic earth movements, (b) lakes formed by vulcanism, (c) lakes formed by erosion, (d) lakes formed by deposition, (e) lakes formed by limestone solution, and (f) lakes formed by organic action including the work of man.

Tectonic movements of the earth's crust may be slow and gradual, creating mountain systems or subcontinental structural basins over a long period of geologic time, or they may be abrupt in the form of earthquakes like that which created Reelfoot Lake, or the one that in 1959 triggered the landslide which partially blocked the Madison River valley in southwestern Montana and thereby gave birth to Earthquake Lake. Some of the largest lakes on the globe were created entirely or in large part by subsidence, filling down-folded structural basins or down-thrown blocks of the earth's crust along or between fault lines. Largest of them all is the Caspian Sea, more than five times the size of Lake Superior, its nearest rival in surface area. The Caspian Sea, once directly connected with the Black Sea and the Mediterranean, occupies part of a great tectonic basin between the Alpine mountain system of Eurasia and the rigid plateau-block which is Africa. Block faulting and graben creation are responsible for the basins of Lake Superior and Lake Tahoe, the great rift valley lakes of East Africa (Victoria, Tanganyika, Nyasa, and others); lakes Väner and Vätter in Sweden, and the Dead Sea on the border of Jordan and Israel. Shallow lakes occupying broader structural depressions, in what were called endoreic regions in Chapter 5, would include Great Salt Lake in Utah, Lake Chad on the southern edge of the Sahara, Lake Titicaca on the Andean plateau in South America, and the Aral Sea and Lake Balkhash in the Soviet Union.

Vulcanism, the movement of molten rock within or at the surface of the earth, is responsible for lake formation in at least two different ways. One is through the collapse of the center of a volcanic cone, creating an undrained depression often but not always roundish in shape, in which a so-called *crater lake* then forms. Such a crater lake might or might not have an outlet, depending upon the balance between runoff and evaporation plus seepage. To most Americans, Crater Lake, Oregon, 2,000 feet deep and more than 6,000 feet above sea level, with a rim that rises another 2,000 feet, is by far the best known example. One of the largest crater lakes in the world is Lake Toba in northern Sumatra, which occupies a caldera basin 750 square miles in extent. This one also has steep 2,000-foot walls surrounding it except where the Soengai Aasahan breaks through in a gorge and tumbles from the rainy heights down to the Strait of Malacca, at one point in a waterfall of 443 feet.[1] Other examples of crater lakes are Maren on the Eifel plateau of West Germany, Bolsena in central Italy, Oskjuvatn in Iceland, and Keloed on the island of Java. Keloed has erupted six times since 1811, temporarily refilling the crater and causing disastrous flooding by the displaced water.

Another type of lake caused by volcanic action is that which forms behind a *lava flow* when such a flow enters a river valley. It was a lava flow into the Jordan valley which created the Sea of Galilee. Lake Kivu in the rift basin of East Africa is a lake of this type as is Lake Van in eastern Turkey. Lake Van, at about 5,600 feet above sea level, was formed by a lava flow from Mount Nemrut into the valley of an upper tributary of the Euphrates. At a certain stage in its development, Great Salt Lake probably had an outlet northward to the Snake River but, if it did, this outlet was later blocked by the great lava flows which formed the Columbia Plateau.

Lakes formed by erosion may be the result of either wind erosion or erosion by ice. In desert regions wind erosion occasionally proceeds all the way down to the water table, creating a shallow lake or swamp. Lakes in the Seistan depression of Iran, the Shotts lowlands on the northern margin of the Sahara in Algeria and Tunisia, and scattered depressions in the desert of Australia are thought to occupy basins which have been deepened by wind erosion. On a smaller scale, the shallow basins in the arid and semiarid Southwest of the United States known as *playas* and *buffalo wallows* may have a similar origin.

Glaciers cause lakes by both erosional and depositional processes. Erosional lakes created by moving ice would include *cirque* or *tarn lakes* in the high-level *catchment basins* where the glaciers are formed and also those lakes farther down the valley caused by gouging or over-

[1]Monkhouse, *op. cit.*, p. 221. This book includes an excellent summary of lake types, with several maps and numerous regional examples.

deepening of the valley floor. Most of the great lakes bordering the Alps of central Europe (Maggiore, Como, Lucerne, Zurich) and the parallel lakes of Norrland province in Sweden were formed, at least in part, by glacial gouging of the bedrock on the floor and sides of a valley. The same is true of many of the lakes in Finland, in the Lake District of northwestern England, in the Finger Lakes region of New York state, in the Canadian Shield of eastern Canada, and in the mountain valleys of Alberta and British Columbia.

Lake basins of depositional origin are likewise mostly associated with glacial action. Except in very high latitudes where they may reach the sea, valley glaciers descending from mountains or high plateaus reach a point where melting is equal to the ice movement, and the *ice front* becomes more or less stationary. At such a point morainic material accumulates on a large scale, and only some of the finer material is carried on down the valley by water. As soon as the ice front retreats, due to a change in climate or any other reason, the *end moraine* becomes a dam and a lake forms behind it. Many of the lakes referred to in the preceding paragraph as the result of ice gouging into the bedrock are much larger than they would be were it not for the presence of a *morainic dam* at the lower end. During the Pleistocene period the mountain glaciers of the northern hemisphere were all very much larger than they are now, and there may no longer be even a remnant of the great glacier responsible for a specific Alpine lake. Both in North America and Eurasia the Pleistocene ice caps extended far out onto the plains, disturbing the mantle rock in massive fashion, pushing rivers out of their courses, and leaving behind tens of feet or even hundreds of feet of debris known as *glacial drift* or *till.* Thousands of hollows in this debris are occupied by shallow lakes, some of them of considerable size.

A special type of lake associated with glacial deposition is the *kettle lake,* occupying a spot where a large ice block was formerly buried in glacial drift. Eastern Wisconsin has many small lakes of this type.

Another special type of lake, usually a very short-lived one, is formed when a large snow slide or avalanche enters a mountain valley. What is believed to have been the largest lake of this type ever formed is no longer in existence. It developed during the last inter-glacial stage of the Ice Age in Europe. The avalanche was caused by the breaking away of more than two cubic miles of a soft limestone formation near Flims, above Chur in the Upper Rhine valley of Switzerland. The limestone rubble spread forward for six miles and created a dam 2,000 feet thick, the remnants of which are still in place. It is thought that the resultant lake survived for many years before the dam broke and the lake was released in a mighty flood.[2]

---

[2]Kuenen, *op. cit.,* pp. 232-234.

Ice itself may form the barrier which creates a lake. Quite a number of such lakes are formed in side valleys draining toward the Vatnajökull ice sheet in Iceland, the big valley glaciers of Alaska, and the margin of the Greenland ice sheet. During the retreat stage of the Pleistocene ice cap in North America, the blockage of the previous drainage outlets to Hudson Bay and the Gulf of St. Lawrence led to the creation of enormous *ice front lakes*. Modern Lake Winnipeg, large as it is, is only a small remnant of Ice Age Lake Agassiz, which for a time overflowed southeastward into the Mississippi River basin. Lakes Erie and Ontario are likewise remnants of a *pro-glacial lake* of much larger size, the lacustrine sediments of which form the fertile flatlands of northern Ohio and Indiana.[3]

On a much smaller scale, lakes in basins of depositional origin would include ox-bow lakes formed through the silting shut of cut-off meanders and other abandoned river channels on floodplains and deltas, lakes, formed behind offshore bars of coastal deposition, and similar lakes or *lagoons* formed behind coastal sand dunes. The coast of France south of the Gironde estuary, the western coast of Denmark, and the coast of East Germany and Poland have many shallow lakes of the lagoon type. Similar ones occur in many places on the coast of Oregon and along the Gulf of Mexico and Atlantic coasts of the United States. The eastern coast of Mexico has many more, and a particularly large one is the Lagoa dos Patos south of Porto Alegre on the coast of Brazil.

Lakes formed by limestone solution are numerous in the Florida peninsula where the largest of them, Lake Okeechobee, has a surface area of 700 square miles about 12-15 feet above sea level. Some of these fresh-water lakes in limestone regions have a surface outlet, but many do not. Solution lakes of the Florida type are present in a number of other *karst regions* in the world, including Ireland, the Jura Mountains, and the Dinaric Alps. Many of the lakes on the floors of the Yugoslav limestone depressions are ephemeral ones, although Lake Shkodra (Skadar) on the Yugoslav-Albanian frontier is a large, permanent one. Settling in consequence of the solution of underground beds of rock salt may be responsible for a few shallow lake basins in Cheshire in western England.

Organic action is responsible for a variety of lakes, some of them very large, if those built by man are to be included. Low-gradient drainage lines in humid regions sometimes become choked by vegetation, causing the formation of shallow lakes. The growth of sedges,

---

[3]Wallace W. Atwood, *The Physiographic Provinces of North America*, Boston: Ginn and Company, 1940, pp. 213-221. The story of the evolution of lake basins and lakes in this area is told in great detail in Frank Leverett and F. B. Taylor, *The Pleistocene of Indiana and Michigan and the History of the Great Lakes*, U.S. Geological Survey Monograph 53, Washington: Government Printing Office, 1915.

rushes, water lilies, and other aquatic plants and their incomplete decay under high water-table conditions, often creates an accumulation of *peat* in poorly drained areas. Pools and small lakes form readily on such boggy moorlands, and the removal of peat by man or by accidental burning tends to enlarge them. On the Lüneburg Heath (German: Lüneburger Heide) of West Germany, the Bourtanger Moor along the German-Dutch border, the Hautes Fagnes of the Belgian Ardennes, the uplands of the British Isles, and almost anywhere in the Arctic *tundra* are many such shallow lakes of vegetative origin. Another type of vegetation-induced lake, according to A. C. Veatch, was that developed on valley lowlands tributary to the lower Red River in Louisiana as a result of silted-in log jams left by the main stream in a prehistoric flood stage. Eventually a sequence of such lakes developed progressively farther upstream until during the last century they were blasted and cleared out, leaving some very fertile *lacustrine flats* available for agricutural use.[4] Somewhat similar to these log-jam lakes, but on a miniature scale, are the dams of sticks and brush built by beavers across many small rivers.

Man-made lakes of various sizes, thousands of them over the world if the small ones are counted, have been built as municipal and community water supplies, millponds for factories, farm ponds to check gullies and water livestock, irrigation reservoirs, hydroelectric projects, navigation channels, wildlife habitats, and recreational centers. There are large ones in all of the continents except Antarctica. Only a few of the major projects can be mentioned here. In the mountains of southeastern Australia, the Snowy Mountains scheme is a complex system of multi-purpose lakes of highly ingenious design. Mention has already been made in Chapter 5 of some of the huge developments on the rivers of the U.S.S.R. The Kakhovka, Zaporozhe, and Kremenchug reservoirs on the Dnieper River; the Tsimlyanskaya reservoir on the Don; the Volgograd, Kuybyshev, Gorki (Gor'kiy) and Rybinsk reservoirs on the Volga are among the largest in European Russia. There are other large ones in the Transcaucasus, such as the Mingechaur reservoir on the Kura River, and those on several other rivers in the Asiatic portion of the country. India, Pakistan, China, and Japan all have large water storage reservoirs serving a variety of purposes, but especially supporting irrigation schemes.

Africa now has several man-made lakes of impressive size completed or in process of construction. The four largest of them (Lake Nasser behind the Aswan High Dam in the United Arab Republic, Lake Volta

[4]Kuenen, *op. cit.*, pp. 237-239, based on A. C. Veatch, *Geology and Underground Water Resources of Northern Louisiana and Southern Arkansas,* U.S. Geological Survey Professional Paper, No. 46, Washington: Government Printing Office, 1906.

FIGURE 6.1 A Reservoir in An Arid Region. Courtesy U.S. Bureau of Reclamation. Boating and other forms of recreation are available on Lake Powell above the Glen Canyon Dam on the Colorado River. The dam is in Arizona, but most of the reservoir is in Utah.

on the Volta River in Ghana, Kainji Lake on the Niger River in Nigeria, and Lake Kariba on the Zambezi River between Zambia and Rhodesia) are together inundating over 7,000 square miles. Between 200,000 and 300,000 people are having to be relocated from the four reservoir sites.[5]

Western and central Europe have numerous reservoirs built for hydroelectric power and other purposes, but most of them are smaller than the Russian and African ones just mentioned. Impressive in both concept and design are the Dutch land reclamation, flood control, and water supply schemes being developed on the basin of the former Zuider Zee. Behind the 18-mile dike, finished in 1932, which effectively shuts out the North Sea but which gives access to it through locks, there is now a fresh-water lake, the Ijsselmeer, bordered by diked and pumped polderlands with attractive new farmsteads, villages, and towns. Equally bold in design is the Delta Plan, begun after the

[5]R. H. Lowe-McConnell, ed., *Man-Made Lakes,* Proceedings of a Symposium held at the Royal Geographical Society, London, on 30 September and 1 October, 1965, published for the Institute of Biology by the Academic Press, London and New York, 1966, p. ix.

FIGURE 6.2 A High Level Reservoir. Courtesy U.S. Bureau of Reclamation. Jackson Lake, elevation about 6,750 feet, at the base of the Grand Tetons in western Wyoming, is a water storage reservoir in a scenic setting, built more than fifty years ago.

disastrous flood of February, 1953, which in the course of a construction period estimated at 20 years will convert several arms of the sea on the compound estuary of the Rhine-Meuse-Scheldt (Dutch: Rijn-Maas-Schelde) delta into freshwater reservoirs.[6]

In the United States numerous man-made reservoirs, some of very large size, have been built by the federal government under terms of the Reclamation Act of 1902, the Tennessee Valley Authority Act of 1933, the Flood Control Act of 1936, and other federal and state legislation. On the Colorado River, Lake Mead above Boulder Dam, Lake Powell above the Glen Canyon Dam, and the Flaming Gorge Reservoir on the Green River tributary are the principal completed projects. In the Pacific Northwest, the Bonneville Dam and the Grand Coulee dam on the Columbia River were finished early in World War II. Franklin D. Roosevelt Lake above Grand Coulee extends up the valley from Grand Coulee for 150 miles or almost to the Canadian border. Several other reservoirs have been completed on the Columbia since the war, and the output of hydroelectric power from that coal- and

[6]K. F. Vaas, "Lakes in Dutch Reclamation Schemes," in Lowe-McConnell, ed., *Man-Made Lakes, op. cit.*, pp. 119-128.

oil-poor corner of the country is now most impressive. A series of huge reservoirs has been developed on each of the Missouri and Arkansas river systems, with a navigation component in addition to flood control and power functions. By 1963, thirty years after the passage of the enabling legislation, the Tennessee Valley Authority controlled a total of 32 reservoirs on the Tennessee River and its principal tributaries.[7] Water development on the river systems of the United States has made a giant stride in the decades of the twentieth century, and the new man-made lakes are a prominent feature of the geographic landscape in many parts of the country.

Physical, Chemical, and Biological Characteristics of Lakes

Lakes, like oceans and rivers, have special physical, chemical, and biological characteristics. In their physical characteristics they differ in depth, surface area, circulation, turbidity, temperature, and a number of other conditions. While pure distilled water is H_2O, a liquid consisting of molecules with two atoms of hydrogen and one atom of oxygen, water as found in nature includes many other substances. Lake water may ordinarily be expected to include a certain amount of carbon, free oxygen, magnesium, manganese, iron, silicon, and other elements. The processes of photosynthesis, organic growth, and decay going on in lakes rearrange and redistribute the chemical elements present, often achieving a kind of cyclical chemical balance, with changes during the year more or less paralleling the physical changes which occur. Since organic life, both plant and animal, has evolved in such fashion as to be finely tuned to physical and chemical conditions of the habitat, a certain ecologic pattern tends to develop in each lake or pond. An externally induced change in the prevailing physical or chemical conditions in a body of water of any size will have immediate repercussions on its habitat quality and, hence, on its biotic content and composition.

The term limnology is applied to the study of the geographical, physical, chemical, and biological features of fresh water, especially fresh-water lakes. Scientists who call themselves limnologists are almost always specialists in biology, so it is the plant and animal life of lakes and the physical and chemical influences on this organic life which tend to dominate the discipline.[8] Space is available here only to men-

---

[7]Smith, *op. cit.*, p. 261.

[8]There is an extensive literature available in the field of limnology. George E. Hutchinson, *A Treatise on Limnology*, New York: John Wiley and Sons, Inc., 1957, 1966, 2 vols., is a good recent survey. A somewhat more popular, but still scholarly study is Robert E. Coker, *Streams, Lakes, Ponds*, Chapel Hill: University of North Carolina Press, 1954.

tion a few of the special concerns of limnologists, particularly as they relate to the activities of man.

The process of photosynthesis—literally "putting together with the use of sunlight"—involves the linking of molecules of water and carbon dioxide to form sugar. Sunlight at certain wave lengths is the energy behind the linking process, and plant chlorophyll is the agent that can apply the link. Since the formation of sugar is basic to all organic production, it is obvious that the availability of light in water is crucial. While the amount of light varies basically with the season in a specific latitude, the amount of effective light also varies with the turbidity or cloudiness of the water. The prevention of abnormal turbidity requires some control of the quantity of clay and other solid particles entering a lake and obviously of the kind of organic waste materials allowed to go into it. Soil erosion alone, however, can be a prime cause of stream and lake pollution.

Free oxygen is released by aquatic as well as by other plants in the process of photosynthesis. This oxygen is then used in the respiration of plants and animals and in the decomposition of dead organic matter. Most of the decay producing bacteria and fungi are aerobic; that is, they are dependent upon free oxygen. When the aerobic agents of decay are unable to function properly it is the anaerobic agents which take over, and the result is a release into the water of undesirable gases like methane, hydrogen sulfide, ammonia, and carbon monoxide. Since these gases are deleterious to most forms of organic life, it becomes extremely important to maintain an appropriate load of free oxygen in the water.

Catastrophic damage to the desirable plant and animal life of a lake can result from the dumping of excessive quantities of raw or partially digested sewage wastes. Industrial effluents that are particularly likely to be harmful include materials that contain toxic substances like sulfuric acid, lead, and phenol; residues that discolor the water or cause unpleasant odor or taste; substances which decompose very slowly like cellulose, ligneous matter, and oils; and any effluent which is bulky enough and hot enough to markedly raise the temperature of the water. The disposal of atomic wastes creates some particularly difficult problems because of the tendency of certain organisms to concentrate radioactivity within their bodies.

Professional conservationists and many other persons are appalled at the damage which has already been done to some of the once attractive and productive rivers, lakes, and ocean margins of the United States. Rivers like the Cuyahoga, the Mahoning, and the Hudson are among the glaring examples. Lake Erie has already been called a "dead lake," and portions of the other Great Lakes, especially the more southerly ones, have badly deteriorated.

Special Characteristics of Selected Lakes

Some short descriptive sketches of individual lakes are introduced at this point to illustrate the variety of physical and human geographic features they display in different parts of the world. The lakes considered here are all large ones, but the largest of all, the Caspian Sea, which is really a detached and slowly dwindling projection of the Mediterranean, is not included.

Lake Baikal in southeastern Siberia, has a surface area of about 12,000 square miles at an elevation of approximately 1,500 feet above sea level. It has somewhat the shape of a new moon, with a length of 395 miles and a maximum width of 49 miles. It is a very old lake in the geologic sense, probably formed in the Jurassic period, but it has unusual survival capacity because of its great depth. The deepest soundings reach a depth of approximately 5,300 feet, or about 3,800 feet below sea level. It occupies a block-fault depression in a severely faulted plateau region. Because of its unusual depth (average about 2,300 feet), Lake Baikal is considered to be the largest reservoir of fresh water in the world, with a volume possibly surpassing that of the five Great Lakes combined. Steep mountain walls line the shore except for the delta of the Selenga tributary which enters on the east side. The outlet is the Angara River, a tributary of the Yenisei system, on which the huge Bratsk hydroelectric plant is located. Ice freezes more than three feet thick in winter on Lake Baikal, and the use of its surface for winter transportation has been noted. In summer there is steamer service between the small communities on its shore. The lake is situated in a forested region, and timber shipping is a major local activity. Many unique types of plant and animal life have developed in the lake, even including one species of seal. Because of its size Lake Baikal does not freeze over until December, while in summer it has a pronounced cooling influence on the surrounding region. Although the nearest large city, Irkutsk, is 35 miles down the Angara from the lake, there is enough local industrial activity to cause some concern about pollution.

Lake Geneva, in central Europe, lying partly in Switzerland and partly in France, has a superficial resemblance to Lake Baikal on the map, for it too is crescent-shaped and has one prominent delta formed by its principal tributary, the Rhone. While its surface area is not one-fiftieth that of Lake Baikal (225 square miles) and its maximum depth is only a little over 1,000 feet, Lake Geneva is still a large lake in its region, largest of all the glacial lakes in the Alps. In a basin formed largely by *glacial gouging,* the lake still receives a considerable volume of meltwater from the Rhone Gacier now far up the valley.

Where the Rhone River enters the lake it is a muddy yellow in color, but where it emerges at the farther end, at the city of Geneva, it is crystal clear. An unusual natural movement of the lake is the *seiche,* a tide-like swell thought to be due to differences in air pressure above it, which can bring a rise of two to five feet at Geneva in a matter of minutes. Lake Geneva is in a densely populated part of Europe that has been occupied continuously at least since the time of the prehistoric lake dwellers, the remains of whose houses have been excavated along the shores. Grapevines cover much of the slope on the north side of the basin. Lake Geneva has had steamboat services since 1823, and there are railroads and highways on both sides of it. The region has many historical and literary associations, and large numbers of tourists visit such lake shore centers as Geneva, Lausanne, Montreux, and Evian-les-Bains.

Lake Chad is different. Situated in a shallow endoreic basin of the African plateau, between the Sahara Desert and the savanna region with its summer rains to the south, Chad has an uncertain future. It seems to be a disappearing remnant of a much larger lake which existed in the more humid Pleistocene period. Sand dunes up to a hundred feet in height occupy much of the abandoned lake surface on the northeast. The remaining lake basin is a double one joined by a single channel through which the movement of water is northward. Water in the Chad basins varies in depth from a minimum of zero to ten feet to as much as 25 or even 35 feet following a series of wetter-than-normal seasons in the tributary areas to the southwest and southeast. In a low rainfall period, the deeper northern basin, which receives the overflow from the southern one, may dry up completely. The area of the water surface in Lake Chad varies from less than 4,000 to about 10,000 square miles. The delta of the Chari River, the principal affluent, is advancing into the lake 150-300 feet per year, and sand dunes, blown by strong northeast winds, are encroaching from the north. While a river capture is not imminent and could be prevented by some engineering work, the Chari is in some danger of a river capture by a tributary of the Benue. The lake is slightly salty, especially in the northern portions, as the rivers keep bringing in a little, and the annual water loss by evaporation is estimated at seven feet. Lake Chad was first seen by European white men in 1823, the year Lake Geneva got its first steamer service. The fact that colonial powers thought it might have both economic and strategic significance, and so raced for it, helps explain why the lake is now divided politically between four countries, Chad, Niger, Nigeria, and the Cameroun. Papyrus (a type of sedge), reeds, and other aquatic plants grow profusely, but the economic value of the lake is slight. A number of tribal people (Buduma, Kuri, Kanembu, and others) live on the shores and islands of the lake, making boats from bundles of reeds, fishing, and raising cattle. A plan is under way to create some agricultural polders along the shore where the soil is suitable.

Lake Titicaca, one of the largest lakes of South America, is situated in the Altiplano structural basin between two ranges of the high Andes. The lake surface, at nearly 12,500 feet above sea level, covers an area of more than 3,000 square miles. The maximum depth is more than 1,000 feet. Like Lake Chad, it is divided nearly in two by a strait, but here there is no danger of either basin going dry. Tributaries enter the lake at many points, some of them fed by melting snow and ice in the mountains which tower several thousand feet above the lake surface. Even though it is in an endoreic region, the water of Lake Titicaca remains fresh, for it has an outlet via the Desaguadero River southeastward to Lake Poopo where the salt is captured and held. The lake straddles the boundary of Bolivia and Peru, and there is a steamer service connecting the Peruvian rail terminal at Puno with the Bolivian terminal at Guaqui. In spite of the high, cold, and quite unpleasant location, the Titicaca lake basin seems to have been occupied very early. The Aymara Indians, whose ancestors were conquered by the Incas and then by the Spanish, retain their language and, to a considerable extent, their ancient way of life. They catch fish in the lake and plant such crops as potatoes, quinoa, barley, and maize on terraces along the shore. Their way of life is a hard one, but they survive in spite of the altitude and poverty of natural resources.

Lake Maracaibo is another lake associated with the Andes mountain system of South America, but one which is a contrast in almost every way with Lake Titicaca. It is located at sea level in western Venezuela, between two northern spurs of the Andean system, the Sierra Nevada de Merida and the Sierra de Perijá. This lake, of about 5,000 square miles area, gets its water not from melting snow and ice but from tropical rains, totaling 70 inches or more a year, which fall on the lake itself and the surrounding slopes. The southern or inner end of the lake is fresh, while the northern end is brackish due to the daily invasion of the tide over the shallow sill separating the lake from the Golfo de Venezuela and the Caribbean Sea. There is a large island in the southern part of the lake and a considerable area of swampland is located along its shores. The mountain slopes are heavily forested, but there is a little farming between the mountains and the lake. The basin is hot, humid, and from all reports an unpleasant place to live.

Spanish explorers on entering the Maracaibo basin in 1499 found Indians living in thatched huts on stilts out over the water. They were reminded of Venice and called the land "Little Venice" or Venezuela.

The city of Maracaibo at the outlet of the lake was a coffee-exporting town of mostly Negroes and Indians when oil was discovered in the basin in 1917. As the oil boom developed, thousands of derricks were erected on shore and far out over the surface of the lake. Hundreds of thousands of people were drawn to the basin by the new oil wealth, and the city of Maracaibo has become not only the principal port of Venezuela but a modern urban center of about half a million people. For forty years after the oil field was developed, only shallow-draft vessels could get over the sill at the north end of the lake, and the cargoes had to be reloaded onto tankers at the port of Maracaibo. In 1957 work was completed on a dredged channel 35 feet deep and 21 miles long which permits 33,000-ton tankers to come directly into the lake and thus avoid trans-shipment.

Great Salt Lake in Utah, at a mean elevation of about 4,200 feet above sea level, bears some resemblance to Lake Chad in Africa in that it is a shallow, shrunken remnant of a much larger lake which formerly occupied its basin. Pleistocene Lake Bonneville, the terraces of which are clearly visible on the surrounding mountains, was almost a thousand feet deep and extended far northward, westward, and southward into Idaho, Nevada, and other parts of Utah. With a drier climate, heavy evaporation, and diversion of some of the former river sources into irrigation systems, the present lake has dwindled to an average depth of perhaps 13 feet (maximum, 35 feet) and a surface area varying in the past century from a maximum of 2,300 square miles in 1873 to a minimum of 1,078 square miles in 1940.[9] The vast salt flats which border the lake have been worked commercially, and the salt supply here is almost unlimited. The salt content of the remaining water in the lake varies somewhat depending upon the lake stage, but in the mid-twentieth century it was holding near 25 percent, slightly under the percentage at which the sodium chloride and sodium sulfate begin to precipitate. Only a few species of small animals, including one type of shrimp, can survive in such saline water, but numerous water birds (pelicans, blue herons, cormorants, terns, and gulls) nest and find their living on the islands and along the shore. The buoyancy of the water, which is seven times as salty as the ocean, has amused many bathers, but the shallowness of the lake and the unattractiveness of the shore have left Great Salt Lake with little recreational value. The lake was discovered by fur traders about 1825 and explored in more detail in the 1840-50 period. Mormon settlers came over the Wasatch Mountains into the eastern part of the drainage basin beginning in 1847. Their followers and descendants have developed an attractive

[9]R. Kay Gresswell and Anthony Huxley, eds., *Standard Encyclopedia of the World's Rivers and Lakes,* New York: G. P. Putnam's Sons, 1965, pp. 108-109.

oasis based on irrigation which reaches from north of the Idaho-Utah line to central Utah and beyond. The golden spike driven by President Grant in 1869 at Promontory Point, on a peninsula which extends into Great Salt Lake from the north, completed the first transcontinental railroad.

Lake Michigan, largest lake located wholly within the United States, has a present surface area of about 22,400 square miles, a maximum depth of over 900 feet, and an average depth of some 275 feet. It came into being during the Pleistocene Ice Age, along with the other Great Lakes, as a result of a combination of tectonic depression and glacial gouging in broad, pre-glacial river valleys. It appears that at one stage a great lobe of the continental ice sheet occupied the approximate present basin of Lake Michigan. During part of the Ice Age, the proglacial lakes at the southern margin of the ice cap drained southward into the Mississippi River and Hudson River systems, but later a rise in the land level and melting of the ice to the north made the present St. Lawrence outlet available. The drainage area of Lake Michigan today is curiously small, with a land surface little more than twice the extent of the lake itself. Water enters Lake Michigan and Lake Huron, which are on the same level (581 feet mean elevation), from Lake Superior via the St. Mary's River. Since the Great Lakes system is all interconnected, and rainfall over the basin, generally 25 to 40 inches per year, is quite reliable, the level of Lake Michigan ordinarily varies only a foot or two in the course of a year. The greatest change of level along the shore, up to six feet or so, comes during severe windstorms. Most of the lake shore is low and sandy, with high dunes in places along the east side. Ice can interfere with shipping on the lake during about four months each winter.

French explorers first entered Lake Michigan in the 1630's, but the vicinity of the lake remained largely Indian country for another two centuries. The site of Chicago was platted in 1830 at the south end of the lake, and the new town grew with phenomenal rapidity to become the second largest city of the nation within less than a century. In the latter part of the nineteenth century it became a major transportation terminal, with huge meat-packing plants and other diversified industries. Because of the growing midwestern markets and the suitability of the location for bringing together iron ore and coal, Andrew Carnegie in 1906 selected the site of Gary, Indiana, on the lake shore about 30 miles east of Chicago, for a huge iron and steel plant. The steel industry at the southeast margin of the lake, the brewing industry on the wets side at Milwaukee, and a host of other industrial and commercial developments all around the lake have brought several million people into the area. The Indiana and Illinois portions of the lake shore and the Wisconsin shore to and beyond Milwaukee are almost a continuous urban area. Pollution of the lake water near the shore and the decline of the fishing industry are not surprising. By agreement with Canada, some Lake Michigan water is used to flush Chicago's processed sewage through the drainage canal on the old Chicago River channel into the Des Plaines River at Joliet and from there into the Illinois-Mississippi system. A federally assisted Great Lakes-to-Mississippi River waterway with a minimum channel depth of nine feet was opened in 1933.

Economic Significance of Lakes

Much of the general economic significance of lakes has been brought out in the discussion of particular examples, but a brief summary seems in order. Some lakes, it must be admitted, have little or no economic value for man but they do, of course, support other organisms, even the bitterly salty water of Great Salt Lake.

Lakes, like the oceans and bordering seas, only on a smaller scale, afford free or low-cost navigation routes. In earlier times, explorers moved into the interior of the continents with much greater ease when they could follow calm stretches of river courses or inland lakes. Portages between lakes and river basins acquired great significance and often saw the early rise of military forts, followed by towns. The exploration stories of North America, Russia, Latin America, and Africa show many parallels, even in very different types of terrain, climate, and vegetation. In regions with cold enough winters, the rivers freeze for a period of months, and the ice surface can then become the roadway.

In many areas without satisfactory alternate land routes, either road or rail, communities along the shores of a lake have found ready means of communication. Sailing-boat service is available on Lake Van, 5,640 feet above sea level on the Armenian plateau of Turkey, and motor launches ply regularly on the Issy-Kul, 5,400 feet high in the Tien Shan mountains of the Kirghiz S.S.R. Passenger schedules are maintained on some of the Finnish and Swedish lakes as well as on the larger Alpine lakes of central Europe. On Lake Victoria and some of the other big African lakes, both passengers and freight move with regularity. Similar service is available in the summer season on the Mackenzie River and Great Slave Lake in western Canada. Automobile-carrying passenger services operate across Lake Michigan, but it is freight traffic which overwhelmingly dominates movement on the Great Lakes. Tens of millions of tons of iron ore, coal, limestone, wheat, and general merchandise are carried each year by shipping lines that operate between lake ports or which come into the lakes through the St. Lawrence portion of the Seaway.

Water supply is often an economic role of lakes. The Great Lakes have long supplied the domestic and industrial water needs of Chicago, Milwaukee, Duluth, Toledo, Cleveland, Buffalo, and other major cities on their shores. The availability of virtually unlimited quantities of fresh water has been an important factor in the growth capability of these centers. Many smaller cities in the glaciated regions of northeastern United States and southeastern Canada secure their water from glacial lakes of lesser size. Syracuse, New York, for example, depends for its water supply on Lake Skaneateles, one of the Finger Lakes. In the western part of the country, Los Angeles gets a sizable fraction of its water supply from the Owens Valley before its runoff reaches brackish Owens Lake.

To an increasing extent, American cities and metropolitan centers of other countries are turning to artificial lakes, usually called *reservoirs,* for their water supply. Some of these reservoirs are simply enlarged or controlled natural lakes, but many of them are completely man-made. Boston looks not only to nearby Framingham Reservoir and Wachusett Reservoir near Worcester but also to Quabbin Reservoir more than 60 miles away in the Connecticut River watershed. To serve the New York metropolitan area, the Croton River, which enters the Hudson above Ossining, now supplies twelve storage reservoirs and six impounded lakes with a total capacity of more than 100 billion gallons. Smaller cities too have their water supply reservoirs. Ogden, Utah, secures water from the Strawberry Reservoir high in the Wasatch Mountains. The writer's home town, Norman, Oklahoma, now draws most of its water from Lake Thunderbird, a reservoir on Little River, a few miles east of the city, built by the U.S. Bureau of Reclamation.

Mineral extraction from lake water is a rather minor commercial activity, but here and there it is of some local importance. Common salt has been recovered from brackish lakes and dried-up lake beds in many countries, including the United States. Ever since the period of the British mandate in Palestine, water from the Dead Sea has supported a potash recovery operation. Searles Lake, in southern California's San Bernardino County, has been a commercial source of potash, bromine, boron, and other minerals. Lake water worth processing for minerals is not likely to have much, if any, other value.

Fishing was a valuable source of food for lake dwellers who lived along the shore or built houses on stilts out over the water thousands of years ago. In the slightly brackish waters of Lake Chad, the crater lakes of Java, the quiet ponds in the *bayous* of Louisiana, and the rock-rimmed lakes of the Laurentian Shield, fishermen are out with their hooks, lines, and nets for very practical subsistence reasons.

Sport fishing engages the interest of millions of people in the mid-latitude industrial countries, and if the worth of the fish caught is sometimes not very impressive, the value of the fishing gear, boats, and camping equipment sold is tremendous. Small man-made ponds for raising fish are common in China, Japan, and many European countries. Hundreds of fish hatcheries to stock natural and artificial lakes are operated by public authorities in the United States.

Lakes, like rivers, may have special political significance. Many of them are divided by invisible but still meaningful political boundaries: to mention a few, Lakes Nyasa, Tanganyika, Victoria, and Chad in Africa; Lake Khanka on the Russian-Chinese border in the Far East; Lakes Shkodra, Constance, and Geneva in Europe; Lake Titicaca in South America; Lake of the Woods and all of the Great Lakes except Lake Michigan in North America. Approval for the construction of artificial lakes on international rivers, like the one at Vianden on the Luxembourg-German border or Falcon Reservoir on the United States-Mexican border, has called for delicate and sometimes prolonged international diplomacy. Pollution prevention, consumptive-use withdrawals, fishing and shipping regulations, control of smuggling, and military maneuvers can all pose problems in the utilization of international lakes.

Swamps

Swamps (or *marshes*) are close relatives of lakes and commonly reflect a dying stage in lake history. The term "disappearing lake" has been applied, especially in the history of exploration, to lake basins where the extent or very presence of a lake depends on the time of year or the stage in a climatic cycle of several years duration. In the desert regions of Utah, Nevada, and southeastern California many lake basins which formerly held water are now barren salt plains or "alkali flats." Sevier Lake in Utah and Carson Sink in Nevada are basins which are dry or nearly dry most of the time but occasionally catch enough runoff to be sizable bodies of water. Lop Nor was the disappearing lake in Mongolia which intrigued Marco Polo and later travelers. The mysterious Lake Torrens, which shows on some early maps of South Australia as extending 400 miles inland from Spencer Gulf, was found by Edward Eyre and other nineteenth-century explorers to be an impossibly desolate expanse of mud covered by white salt crystals and occasionally a vast sheet of green, brackish water. Modern maps of Australia recognize several different "lakes" in this great salt marsh: Torrens, Eyre, Gairdner, Frome, and others. The Okavango Swamp in northern Botswana is a similar natural feature in the Kalahari Desert of South Africa.

While there are numerous swamps or marshes in the world's dry lands where tectonic movements, wind erosion, or other factors are responsible for the basins in which they form, there are, of course, many more in humid regions. Here they are not likely to be saline unless they are tidal or very close to sea level, like the Hackensack Meadows of New Jersey. The Dismal Swamp of Virginia-North Carolina, the Okefenokee Swamp of Georgia-Florida, and the Everglades of Florida are all a few feet above sea level and so are fresh-water swamps. On the Arctic coast of North America and Eurasia there are vast flat lands that are swept over by salt water during times of strong onshore winds. The Mackenzie and Lena river deltas are notable examples of such half land-half sea terrain. Another such area of half land-half sea is the extension of the Volga River delta into the northern end of the Caspian Sea. Channels have had to be dredged through the marshes to permit the boats of the Caspian fishing fleet and other vessels to reach Astrakhan, the major urban center on the delta.

By far the largest fresh-water swamp in the world is that which covers much of the northern half of the Ob River basin between the Ural Mountains and the Central Siberian Highlands. Here is an area of hundreds of thousands of square miles covered by low-grade forest and marsh vegetation where the Ob and its tributaries wander almost without gradient northward toward the Kara Sea. Settlements are few, always small, and practically always on navigable streams. The current Soviet effort to explore this vast swamp for oil and gas and to build pipelines from the producing fields to far off centers of consumption is encountering formidable natural obstacles. Russian petroleum engineers gained valuable experience for this effort during the decade of the 1950's when they were assisting the Chinese government with an oil-drilling program in the marshy Tsaidam basin of western China. American engineers involved in petroleum and sulfur production on the Louisiana coastal marshes are familiar with many of the same technical problems.

Fresh-water marshlands are present on upland plateau and mountain surfaces in many regions, but they are especially widespread in the recently glaciated lowlands of northern Europe and northern North America. If forested, the height and quality of the forest is usually low because the high water table restricts the functional zone of the root systems. Sedges, rushes, and other aquatic plants occupy vast areas adjacent to patches of open water. Numerous rivers have their sources in or flow through such marshes. In the more southerly parts of the glaciated regions, many lakes and marshes have dried up, and their basins have been converted to meadow or crop land. Drainage ditches have been dug and tile drains have been installed in many areas to speed up the natural process. Some of the marshy bottomlands in states like Ohio, Indiana, and Iowa have been turned into black, highly valuable vegetable fields where the "muck" is worth hundreds of dollars per acre.

The Pontine Marshes of southern Italy were converted into new farmland, spotted with bright new towns, in a reclamation operation between the two world wars, somewhat comparable to the drainage of the Fens in England in the seventeenth century or the polderland projects in the Netherlands. The Pripet Marshes in the western part of the U.S.S.R. have at times been used as a system of natural defenses, but they too are now in the early stages of conversion to farm land through a long-range program of ditching, diking, and pumping.

Most areas of swamp and marshland have little present value. They are an obstacle to settlement and transportation routes which might be needed through them. In spite of minor exceptions like the wild rice industry of Minnesota and Wisconsin or the cranberry industry of Cape Cod, their vegetation is not marketable. Although they retain large amounts of usable water, much of it, like that of El Sudd in the Sudan, evaporates before it can be moved to points of possible use. Where rock structures are suitable, seepage from swamps may help to hold up the water table in adjacent areas where the water supply has to be pumped from the ground. Marshlands often have their greatest values as wildlife preserves where water birds can nest and find food and where alligators (if you happen to like alligators) can be preserved for the enjoyment of posterity. They may also be a place of refuge from enemies, a reasonably safe place to operate a moonshine still, or to hide out from the law for those who are sought by police.

Chapter **7**

Water in the Ground

Of the total available supply of fresh water in the earth's hydrosphere, less than one percent is in the rivers and lakes discussed in Chapters 5 and 6. About three-fourths of it (74.7 percent) is locked in the form of ice in the polar ice caps and glaciers, and a little less than one-fourth (24.9 percent) is underground.[1] The water potential in the glacial ice is presently of little use to man in any direct sense. Most of the water currently being used is that in rivers and lakes on the surface, but in many places and circumstances *ground water* use is predominant.

Origin and Nature of Ground Water

A very small fraction of the water in the ground is *connate water,* water retained in rocks since the time of their original formation. A bit more, it appears, is *magmatic water,* released as a result of igneous action in the movement of molten magma underground or its emergence on the surface as lava. Nearly all underground water, however, is *meteoric water,* directly derived from the insoak of rainfall, melting snow and ice, and water bodies on the surface. Some of the water penetrating the ground surface remains underground for only a short period of time as much soil moisture is soon returned to the air by direct evaporation or by the transpiration of plants. When the surface is sufficiently moist, however, and the mantle rock and bedrock will permit, there is a downward percolation to a *zone of saturation* in which all of the rock interstices are filled with water. The top of this saturated zone is the *water table.*

[1]Edward A. Ackerman and George O. G. Lof, *Technology in American Water Development,* Baltimore: The Johns Hopkins Press, for Resources for the Future, Inc., 1959, p. 12.

The term water table, although widely used, is somewhat misleading in that it suggests that the top of the saturated zone is flat. It is not flat, but rather it tends to parallel the ground surface except in a somewhat smoother pattern. The water table is normally somewhat closer to the ground surface beneath valley floors and lower slopes than on hilltops and upper slopes in rolling terrain. In some locations a drilled water well might encounter a *perched water table* somewhere above the true water table because of the intervention of a layer of clay or other impermeable rock above the main porous zone where the regular water table develops. Water naturally soaks in much more readily when the surface soil is loose and open and the subsoil and bedrock are of porous material.

The effective *porosity* of the mantle rock and bedrock varies greatly with the type of material. The quantity of water which can be absorbed in relation to the volume of rock material penetrated is generally between 45 and 60 percent in the case of clay-type, loamy, or sandy soil; 30 to 40 percent in sand; 5 to 20 percent in sandstone, 1 to 8 percent in limestone and dolomite, and 0.3 to 0.8 percent in granite. Most igneous and metamorphic rocks, such as granite, quartzite, schist, and slate, have very little absorptive capacity other than in cracks and joints. At increased depths below the surface, the amount of penetrable space (because of the superincumbent weight) is less and less, reaching a point of inconsequence in most sedimentary rocks below about 5,000 feet and in crystalline rocks below about 500 feet.[2]

Not all ground water is fresh. In northwestern Oklahoma salt-water springs flow from the banks of the Salt Fork of the Arkansas River and an upper tributary of the Cimarron River. At deeper levels the water in saturated zones of sedimentary rock is sometimes heavily charged with salts and represents a repository of ocean water from some earlier geologic period. Oil wells drilled to several thousands of feet depth often encounter zones of salt water. Saline ground water may be present at or near the surface in the vicinity of the ocean margins, especially when there have been excessive withdrawals by pumping of fresh water. Such penetration of surface sediments by ocean water has been a problem in the San Francisco Bay area of California and in the Dutch polderlands immediately behind the coastal dunes and tidal estuaries. Since the saline water is slightly heavier, additional hydrostatic pressure has to be applied to the fresh ground water surface to force the salt water to withdraw.

Ground water is much less likely than surface water to be polluted by organic substances or industrial wastes and is therefore often favored

[2]Daniel W. Mead, *Hydrology: The Fundamental Basis of Hydraulic Engineering*, New York: McGraw-Hill, Inc., 2d ed., 1950, p. 452.

in rural areas or urban communities where purification facilities are not available. Because ground water has been in contact with rocks over a longer period of time than fresh water, it usually contains more dissolved minerals. The most common mineral salts likely to be present are compounds of sodium, calcium, magnesium, and potassium, although iron, boron, and other substances may also be represented. Water which contains more than a certain amount of calcium and magnesium salts is said to be *hard water.* Hardness is ordinarily noted in reference to the amount of calcium carbonate present. Water is soft if it contains 0-60 parts per million of calcium carbonate, moderately hard with 61-120 ppm., hard with 121-180 ppm., and very hard with more than 180 ppm.[3] If there are more than 500 ppm. of dissolved solids, the water is not considered suitable for domestic use, although higher levels can be tolerated. *Soft water* is usually preferred for laundering, washing, and bathing, but it does not necessarily taste as good as water with a higher mineral content. Water that is too hard leaves a scaly deposit on water pipes, boilers, and teakettles, and excessive iron in water causes discoloration and stains on fixtures and cloth. Sophisticated water treatment plants strive for both purity and a reasonable balance between hardness and softness. Some industrial water users require very careful control of the mineral content. Sheep are unusually tolerant of dissolved salts in their drinking water and can use supplies that are twice as saline as can safely be employed in irrigation.

The presence of water underground does not necessarily mean that this water can be withdrawn by pumping. Water as a liquid has a high surface tension and clings tightly to small solid particles such as are present in a clay or shale. It will move upward or downward through such fine materials by capillary action, but it will not concentrate within an open space such as is created at the bottom of a well. The amount of water which can be withdrawn from a given saturated zone depends more on the size than on the total volume of the pore spaces present. The term *aquifer* is applied to a saturated layer of rock, whether consolidated or not, through which water will move readily in response to pumping. Unconsolidated gravels and sands, such as are found in quantity in most glaciated areas and in many river beds, make excellent aquifers. Most sandstones and many limestones, dolomites, and chalks are also suitable. In areas of crystalline bedrock and little or no mantle rock the amount of ground water present is limited, but what there is can often be pumped quite readily. If physical conditions within a given aquifer are suitable, thousands of gallons of water per minute can be withdrawn from a single well over a sus-

[3]U.S. Geological Survey, *A Primer on Ground Water,* by Helene L. Baldwin and C. L. McGuinness, Washington: Government Printing Office, 1963, p. 11.

tained period. In 1951, Harold E. Thomas, of the U.S. Geological Survey staff, published a map of the United States which has been very widely reproduced, showing the general regional adequacy of ground water supplies.[4]

The volume of ground water available in a given aquifer can be calculated approximately by a pattern of borings distributed over its horizontal extent, combined with an evaluation of the percentage of the water present which can be withdrawn by pumping. The thickness of an effective aquifer might vary from a few feet to as much as several hundred feet. It is not unusual for the total available ground water in a given area to be distributed among two or more effective aquifers separated by less porous rock layers from which pumping would be unproductive.

When a particular aquifer is not horizontal, the water in it tends to move slowly down slope from the source area where it is supplied. If farther down the dip the aquifer is capped by an impervious rock formation, the water in it may be under considerable pressure. Should the pressure become great enough, the water may find its way to the surface with force when a well is drilled into it. Ground water confined in this manner is said to be under *artesian pressure* and the flowing well is an *artesian well,* taking its name from the district of Artois in northern France where such wells were already known in the Middle Ages.

In the early twentieth century there were many artesian wells in eastern South Dakota, producing from the Dakota sandstone aquifer which outcrops near the Black Hills and along the eastern front of the Rocky Mountains. Across much of its eastward dip the Dakota formation is capped by the nearly impervious Pierre shale, and when first drilled into some three or four hundred miles east of the source area it released its water with considerable force. Another region which has had a large production of water from artesian wells is the sheep-ranching country of east central Australia, where the water entering the aquifer in the eastern mountains has been held under pressure in similar fashion to that enclosed in the Dakota sandstone. Some of these artesian wells in Australia are more than 3,000 feet deep.

When water is pumped from an underground aquifer its removal results in what is called a *cone of depression.* The water table is lowered somewhat all around the well so that the surface of the water table in the vicinity of the well resembles an inverted cone. The potential yield from a well depends on the amount of lateral water pres-

---

[4]Harold E. Thomas, *The Conservation of Ground Water: A Survey of the Present Ground-Water Situation of the United States,* New York: McGraw-Hill, Inc., (Sponsored by the Conservation Foundation), 1951.

sure tending to eliminate the cone. It is customary in the United States to rate the productivity of ground water wells in gallons per minute under normal pumping conditions. The spacing of wells on a given aquifer must take into account the amount of draw-down in adjacent cones of depression. In general, the more porous the aquifier the more closely the wells can be spaced. In Norman, Oklahoma, prior to the development of a surface source in Lake Thunderbird, municipal water was drawn from the Wellington-Garber sandstone formation in a pattern of wells spaced at half-mile intervals.

It is quite easy to observe the relation between the rate of supply and rate of withdrawal when water is taken from a surface source such as a river, lake, or reservoir. It is much less easy when water is produced from ground water wells. The water table rises and falls in relation to rainy or dry seasons and wet or dry cycles extending over a period of years. Most irrigation farmers and other consumers of ground water have quite inadequate information as to how their rate of use compares with the local rate of ground water recharge. In an effort to evaluate more accurately the adequacy of ground water supplies in relation to withdrawals, the U.S. Geological Survey by 1964 had installed 3,500 observation wells and was planning to increase the number to 7,000 by 1973. Reasonably good descriptive information is available for several of the eastern and south-central states, but precise quantitative information is limited to a few hundred counties scattered over the country. In making its studies the Geological Survey ordinarily works in close cooperation with state and local agencies.[5]

Over most of the United States the surface and ground water supply is adequate for present and short-range future needs, assuming the necessary outlays for storage and distribution systems. In the southwestern part of the country, however, a number of ground water reservoirs are subject to perennial overdraft, and the water table is falling markedly. Ground water mining is being carried on in the rice-growing areas of southwestern Louisiana and eastern Arkansas and in a number of cotton and horticultural regions from west Texas across New Mexico and Arizona to southern and central California. Thomas[6] described the situation as critical in several of these areas of intensive irrigation almost twenty years ago, but the mining goes on and the water tables continue to fall.[7]

---

[5]U.S. Geological Survey, *Long-Range Plan for Resource Surveys, Investigations, and Research Programs of the United States Geological Survey, 1964,* Washington: Government Printing Office, 1964, pp. 42-57.

[6]Thomas, *op. cit.*, pp. 36-77.

[7]*The Nation's Water Resources*: The First National Assessment of the Water Resources Council, Washington: Government Printing Office, 1968.

Artificial recharge of overpumped reservoirs through water spreading on the surface and discharge into specially designed wells has long been practiced in Sweden, Germany, and other European countries. Such efforts have usually been quite local in nature, however, mainly to restore the underground supply of particular cities.[8] Similar recharge techniques have been employed in the American Southwest, as in the San Fernando Valley of California, but the available supply of recharge water is totally inadequate.

Dependence on ground water for domestic and industrial water supplies seems quite feasible for small and medium-sized urban communities located in glaciated regions or in or on the edge of broad river valleys with a floor of coarse sediments. Areas of heavy glacial deposition are particularly fortunate in this regard. In the British Isles, Scandinavia, the Netherlands, Germany, Poland, and the northern half of European Russia there is an abundance of ground water, usually within a short distance of the surface. Very large cities and densely populated areas in general often develop surface water reservoirs to supplement their ground water supplies. It was estimated in 1955 that in the United States as a whole nearly 80 percent of the water obtained from wells came from loose or slightly consolidated sand and gravel deposits; about five percent from limestone, mainly in Florida, Georgia, New Mexico, and Texas; about three percent from sandstone in various parts of the country; and two percent from porous basalts, chiefly in the Columbia Plateau of the Pacific Northwest.[9] An excellent general summary of ground water utilization in the United States was published by the U.S. Geological Survey in 1963.[10]

Relatively little ground water is stored in natural caves. Limestone solution caves in karst regions, of the type mentioned earlier in relation to lost rivers, may have sizable pools of standing water, but these are rarely of a magnitude that would support sustained pumpings.

Springs

Springs are natural flows of water originating underground and emerging where a water table or perched water table intersects the ground surface. River channels are often at or very close to the water table, so that many rivers are spring-fed without that fact being very

---

[8]Oscar E. Meinzer, ed., *Hydrology*, New York: Dover Publications, Inc., 1949, pp. 404-405.

[9]Harold E. Thomas, "Underground Sources of Our Water," in *Water*: The Yearbook of Agriculture, 1955, *op. cit.*, p. 73.

[10]*The Role of Ground Water in the National Water Situation,* by C. L. Mcguinness (with state summaries), U.S. Geological Survey Water-Supply Paper 1800, Washington: Government Printing Office, 1963. 1121 pp. and maps.

obvious to the casual observer. The more typical concept of a spring is that of water bubbling out of a hillside or mountain slope and finding its way (often through a culvert under the highway) to a valley somewhere down below. Many springs have an active flow only during the rainy season and become dormant the rest of the year. The U.S. Geological Survey has mapped and described the larger springs of this country in some detail. A few of them bring to the surface enough water to supply the domestic needs of many thousands of people. Large springs are especially numerous in limestone regions where they may be simply the re-emergence of a lost river, but others are in regions of porous basalt or thick gravel deposits. Numerous springs are present in northern Florida, the Missouri Ozarks, central Texas, central Montana, the Snake River plains of southern Idaho, western Oregon, and northern California. A line of *fontanili,* large springs, has developed along the northern edge of the Po River lowlands in Italy, fed by seepage out of the thick end moraines left by the Alpine glaciers.

Ground water may emerge as springs with a temperature anywhere from just above freezing to boiling. Water pumped to the surface from very deep levels is likely to be warmer than that drawn from nearer the surface. Springs with an unusually high mineral content, which commonly precipitate some of the minerals as they emerge from the ground, are known as *mineral springs.* Frequently such springs emit warm or hot water, in which case they are called *thermal springs.* The heat is generally derived from a mass of very hot magma somewhere beneath the surface in a zone reached by the underground water. There are thousands of such hot springs in the world, most of them in igneous regions of present or past volcanic activity.

Water from some mineral springs has, or is reputed to have, a certain health or curative value. It may be bottled, specially labeled, and widely sold for drinking. Thermal springs in many countries have been developed as tourist attractions featuring mineral baths, where the hotel or a whole cluster of hotels offers fine meals, special concerts, and all sorts of arranged diversions. The city of Bath in England; Spa in Belgium; Karlovy Vary (German: Karlsbad) in Czechoslovakia; Bad Nauheim and Baden-Baden in West Germany, and Vichy and Lourdes in France are well known European centers of this type. In the United States, Saratoga Springs, New York; Warm Springs, Georgia; White Sulphur Springs, West Virginia; and Hot Springs, Arkansas have had a somewhat comparable development.

In some regions of thermal springs the hot water may emerge from such deep-seated and irregular vents that it is heated to or beyond the boiling point and forced out intermittently under considerable steam pressure. Such hot fountains are called *geysers.* The three best known geyser areas in the world are Iceland, Yellowstone National Park in the United States, and the North Island of New Zealand. Some of the thermal water of Iceland is used in a central heating system in Reykjavik, the capital city. In New Zealand and a number of other places the heat in thermal springs is used for the generation of electricity. Hot steam, often mixed with other gases, is vented from the ground in a number of places including Java, the Auvergne region of France, the Phlegraean Fields west of Naples in Italy, and the Valley of Ten Thousand Smokes in Alaska.

Economic Significance of Ground Water

Quantitative data is not available to evaluate with any precision the role of ground water in most countries in the world. Its use is obviously indispensable in those inhabited areas (and they are found in many countries) where little rain falls and there is no reliable surface supply. The Old Testament, whose setting is largely in an arid environment, is replete with references to ground water, springs, and wells. C. F. Tolman[11] and other writers have described the large underground *water tunnels*, or *kanats*, used in Egypt and Persia as early as 800 B.C. In many parts of the Middle East and North Africa there is almost total dependence on dug wells, frequently carved deeply into the local bedrock. Jean Brunhes, writing early in the twentieth century, described in vivid detail the community life of the Mzab oasis settlements in the Algerian Sahara, where men and donkeys labored by the hour to draw water with bucket and rope from dug wells more than 50 meters deep.[12] The "lady at the well" is still a familiar scene in many North African oases. The "old oaken bucket," now little more than a museum relic, once played an important role in the rural and even the urban economy of Europe and North America. The hand-operated lift pump was a welcome invention when it came, as were the more sophisticated vacuum, hydraulic, and turbine pumps which followed. Large powered pumps, driven by electricity, fuel oil, or natural gas, now draw out streams of water four, six, or eight inches in diameter to supply cities and irrigate millions of acres of farm land.

Ground water is used on a large scale for domestic and industrial purposes in western Europe, especially in the glaciated northern countries. Highly sophisticated studies of ground water reserves, aquifer recharge rates, artificial recharge techniques, and underground water

---

[11]C. F. Tolman, *Ground Water,* New York: McGraw-Hill, Inc., 1937.

[12]Jean Brunhes, *Human Geography,* Chicago: Rand McNally and Company, 1920, pp. 432-452. (Translated from French by I. C. LeCompte; edited by Isaiah Bowman and Richard E. Dodge.)

quality are being made in the United Kingdom, France, the Benelux countries, Germany, and Scandinavia. Per capita water consumption in European cities, which before World War II was only about one-fourth that of equivalent American cities, has been increasing rapidly. In Sweden the cities of more than 100,000 population (Stockholm, Göteborg, and Malmö) are largely drawing on surface water supplies; cities of 20,000 to 100,000 are mainly using artificially-infiltrated ground water sources, and cities with fewer than 20,000 are getting most of their water from natural ground water reserves. Many wells in Sweden are drilled deeply into the crystalline bedrock, and such cities as Hälsingborg and Eslöv in Skåne province draw much of their water from such bedrock sources.[13]

In the United States, ground water reservoirs in 1960 supplied a little less than one-fifth of the total surface and ground water withdrawals for all uses. By 1968 the estimate had increased to 22 percent.[14] All 50 states are substantial users of ground water, and ground water use is estimated to have doubled between 1945 and 1960. About 75 percent of American cities derive their water from wells, although most of the larger metropolitan centers depend mainly or exclusively on surface sources.

Irrigation is the largest single user of ground water. California, Texas, and Arizona, in that order, are the leading pumpers. In the southern High Plains of Texas, which was mentioned earlier as one of the regions of serious water mining, the highly productive Ogallala formation once stored nearly 250 million acre-feet of water. The semiarid climate and present topographic situation of this region restrict recharge to less than two percent of the rate of withdrawal. By the end of 1961 almost 50 million acre-feet of water had been pumped from the Ogallala formation, or one-fifth of the amount originally present.[15]

Irrigated acreage in the United States increased from less than eight million acres at the turn of the century to more than 37 million acres in 1964. The 17 states in the western half of the country have less than 30 percent of the nation's estimated water supply, but they account for 84 percent of all water consumption. Almost 90 percent of the water used for irrigation is used in the 17 western states. Irrigation (mainly pump irrigation) has been increasing with special rapidity in the six plains states from Texas to North Dakota. In the 20-year period from 1944 to 1964 the agricultural area under irrigation in these

---

[13]Martin G. Beyer, "Ground Water Production from the Bedrock of Sweden," in E. Eriksson, Y. Gustafsson, and K. Nilsson, eds., *Ground Water Problems*, Proceedings of the International Symposium held in Stockholm, October, 1960, Oxford: Pergamon Press, 1968, pp. 161-162.

[14]*The Nation's Water Resources, op. cit.*, p. 3-2-7.

[15]U.S. Geological Survey, *A Primer on Ground Water, op. cit.*, p. 24.

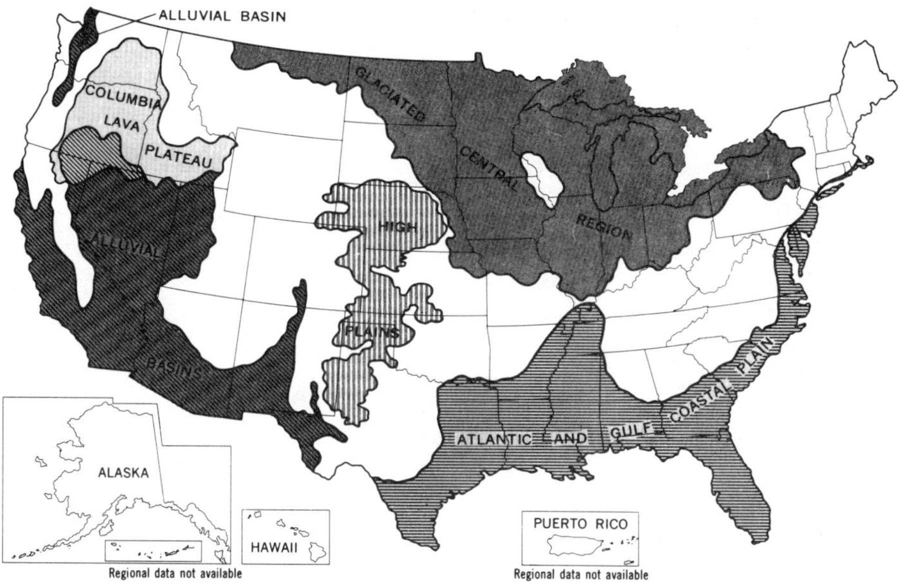

FIGURE 7.1 Major Areas of Potential Ground Water Development in the United States. The areas in white on this map are those where bedrock with little or no potential for large-volume ground water production are near the surface. More precise information is shown on the H. E. Thomas ground water map referred to in the text. From United States Water Resources Council, **The Nation's Water Resources,** p. 3-2-8.

six states increased from 2,126,000 acres to 10,041,000 acres, or by almost five times. In Kansas during this 20-year period the irrigated acreage increased by more than ten times, and in Oklahoma, where there was little irrigation before World War II, by more than 100 times![16] In Nebraska, where gravity irrigation based on Platte River water had been important for several decades, the irrigated acreage expanded by more than three times, mainly on the flat lands between the Platte and Republican Rivers in the south-central part of the state and around the northern and eastern margins of the Sand Hills. The number of irrigation wells in Nebraska increased about 25 percent each year during the decade of the 1950's, and by 1960 there were more than 23,000 in production, drawing several billions of gallons of water a day during the irrigation season.[17]

---

[16]U.S. Department of Agriculture, Economic Research Service, *Major Uses of Land and Water in the United States with Special Reference to Agriculture, Summary for 1964,* Agricultural Economic Report No. 149, Washington: Government Printing Office, 1968, pp. 34-41.

[17]U.S. Geological Survey, *A Primer on Ground Water, op. cit.,* p. 23.

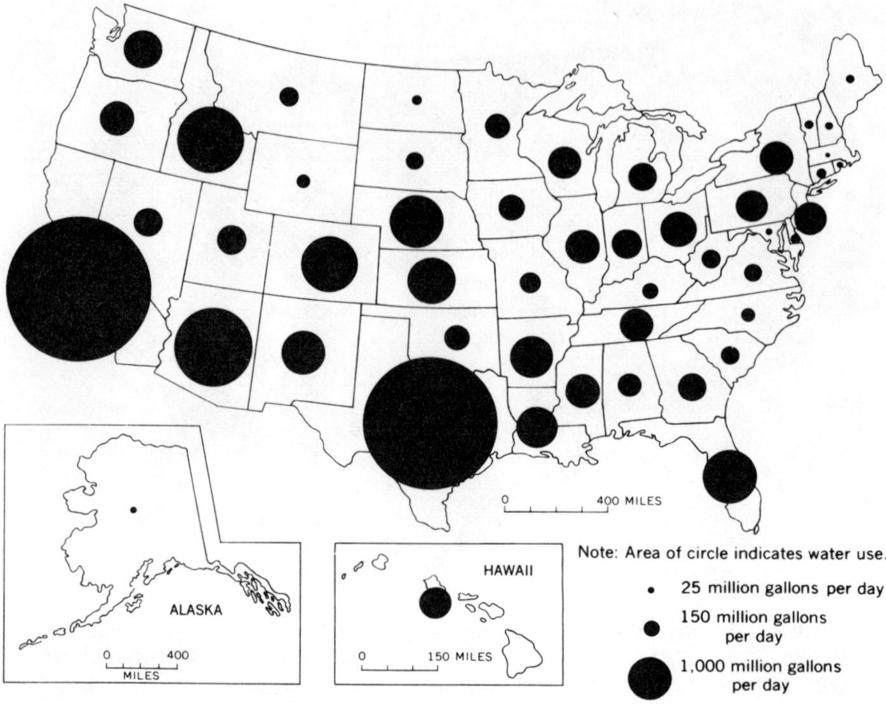

FIGURE 7.2 Total Use of Ground Water in the United States, 1960. In the Great Plains region, especially from Nebraska southward to Texas, there has been a substantial increase in ground water withdrawals since 1960. From United States Geological Survey, **A Primer on Ground Water,** p. 23.

Three relatively new techniques of reducing the loss of water from surface reservoirs, canals, and irrigation ditches are now under experimentation and/or development: (a) coating the water surface of reservoirs with hexadecanol or some other chemical substance to retard evaporation; (b) poisoning or otherwise removing *phreatophytes,* the undesirable aquatic plants which grow in such profusion along rivers, canals, and ditches; and (c) lining earthen ditches with an impervious chemical coating or heavy plastic sheeting to reduce bottom losses from seepage.

While the Great Lakes and other surface water sources meet most of the water demands of the major American cities, the surface supplies of New York City, Los Angeles, San Francisco, Indianapolis, Columbus, Louisville, Pittsburgh, and St. Louis are all supplemented by wells. In the early 1960's, wells supplied all the water for 150 cities with populations of 25,000 or more, including such major metropolitan cen-

FIGURE 7.3 Raised Ditch Irrigation Using Gravity Flow. Courtesy U.S. Soil Conservation Service. This farmer, near Republican City, Nebraska, is using siphon tubes to distribute gravity-flow water from a raised ditch between rows of tall corn.

FIGURE 7.4 Sprinkler Irrigation. Courtesy U.S. Soil Conservation Service. The self-propelled irrigation system shown here on a farm near Page, Nebraska, turns slowly in a huge circle to water newly planted brome grass and alfalfa.

ters as Houston, San Antonio, Wichita, Dayton, and Miami.[18] A convenient summary of water sources supplying each of the 222 standard metropolitan statistical areas (SMSA's) of the country has just been published by the U.S. Geological Survey.[19]

The possibility of systematically replenishing ground water in order to maintain subsurface reserves is now under intensive engineering study in a number of countries. Relatively simple techniques for diverting flood waters away from river channels so that it will have more time to soak in and replenish the ground water have long been in practice in Morocco and Algeria. The qanat system of storing ground water in underground tunnels, still in wide use in Iran, has the obvious advantage of sharply reducing evaporation losses. Much fresh water moves toward the sea in the loose sediments of river channels even when there is no surface flow. Dams can be built across these channels to retard the subsurface flow and make more of it available for pumping. Massawa, on the Red Sea coast of Eritrea, gets its water supply in this fashion, and a similar technique has been employed in other dry areas of Africa and South America.[20]

Specially designed wells for injecting surplus water underground have been tested with some success in various European countries and in the United States. The savings in reduced evaporation from surface reservoirs are obvious when this type of underground storage can be employed. Systematic surface flooding on porous surfaces is being tried in a variety of circumstances in Texas and California as a means of bolstering the ground water supply. A serious complication in the water-flooding technique is finding some way of removing silt from the water before it is put into the seepage basins, because otherwise the silt tends to seal up the flooded surface and reduce its rate of insoak.

Some conservationists have objected strenuously to the drainage of "wetlands" for agricultural use, not only because the practice destroys wildlife habitats but because it reduces the support these wetlands afford to regional ground water supplies. Ground water can be a permanent resource, in that in nearly all areas it is subject to some rate of recharge. If enough information were available, it would at least be theoretically possible for public authorities to meter it and restrict the rate of use to the long-range rate of replacement. The total amount of fresh ground water available is enormous. In the United States the

---

[18]Smith, ed., *op. cit.*, p. 243.

[19]*Water Data for Metropolitan Areas*: A Summary of Data from 222 Areas in the United States, compiled by William J. Schneider, U.S. Geological Survey Water-Supply Paper 1871, Washington: Government Printing Office, 1968.

[20]F. Dixey, "Water Supply, Use and Management," in E. S. Hills, ed., *Arid Lands: A Geographical Appraisal*, London: Methuen & Co., Ltd., in cooperation with UNESCO, 1966, pp. 85-96.

amount has been estimated at the equivalent of about 30 years of precipitation, or some 75 years of the mean surface runoff of U.S. rivers. If used with knowledge and a reasonable measure of restraint, this ground water can be a priceless resource for the future as well as for the present.

Chapter **8**

The Management of Water

Water is a natural resource available (in a general sense) in almost unlimited abundance. For a large fraction of the world's population, however, water is far from being a free or even easily accessible resource. Physical access to it, the legal right to use it, and the actual opportunity to consume it in the quantity and quality desired are all limited. To make the water of the globe of maximum use to man, more information needs to be available concerning its nature, its behavior, its location, and its capacity to satisfy human needs and desires. Much more needs to be known about the way to manage water effectively, to get it to the point of use at minimum social and economic cost and with proper attention to quality. As world population increases, it seems obvious that more rather than fewer administrative regulations and controls will be necessary. Numerous priorities will have to be agreed on, and these priority decisions will have to be made at local, state, national, and international levels.

The management of water at the local level may involve private, corporate, or municipal decisions about digging a well, piping water into a house or factory, installing a pump, damming a creek to make a pond, irrigating a tract of land, or building a water treatment or sewage disposal plant. A village or small town in some parts of the world might settle for a single well or water point to which residents could come with jars or buckets to get their daily requirement for personal needs. In a modern society with more types of demands on the water supply, a small town which happens to be favorably situated over a reliable aquifer with naturally pure water can still get along with a good pump, a tall water-storage tank, and a relatively simple distribution system. Large municipalities, however, are confronted with water management problems of much broader scope; they have to make decisions about storage reservoir sites, alternate sources of supply, access routes for water mains and sewer lines, complex processing equipment, and priorities for various types of users. The challenge of raising funds for public water systems and equitably allocating costs among the users of publicly supplied water is a well known city hall problem.

On the state level in the United States, many decisions concerning water are made by the legislatures and administrative officials. Systems of water law are adopted or developed. Hydrologists, snow measurement experts, hydraulic engineers, water quality specialists, and professional personnel qualified to interpret and enforce water laws are employed. Governors' committees are appointed to help establish water-use priorities and develop long-range state water plans. Interstate compacts, made possible by federal legislation, have to be worked out between adjacent states concerning withdrawals from river flow, lake water, or ground water sources along state boundaries.

The federal government in the United States, having control of navigable streams and interstate commerce and having responsibility for national defense and the promotion of the general welfare, has become involved in a very broad array of water problems. The improvement of seaports and access to them, the deepening and straightening of river channels for water transportation, the building of canals, the control of floods on navigable rivers and their tributaries, the production of electric power, the development of recreation facilities on natural water bodies and reservoirs, the provisioning of water for domestic, industrial, and irrigation use, and the maintenance of public health through the enforcement of minimum water quality standards are some of its preoccupations.

The U.S. Supreme Court has been asked to adjudicate many water rights and water boundary cases. The legal problems concerning interstate transfers of water, water pollution, and artificial weather control (such as rain-making) are intricate and delicate.

There is no National Water Plan, although there are signs of increasing need for one. This nation does now have a national Water Resources Council with broad planning responsibilities, established under the Water Resources Planning Act of 1965. Of even more recent creation is the National Water Commission, which is to "review and advise the nation on the entire range of its water-resource problems."[1] Eventually, it would appear, priorities having to do with access to water, and projects for water development will have to be established in a more deliberate, scientific, economic, and humane manner than they

[1]*The Nation's Water Resources, op. cit.* pp. iv-v.

sometimes seem to be at present.[2] This book, however, is not the proper place to "beef" about the "pork barrel" approach to public improvements!

Water, in certain respects, has been an international problem for a long time. The right to sail the sea and engage in international commerce has been claimed by nations and disputed by nations since ancient times. Pirates and buccaneers have attacked ships at sea, raided port towns, and sailed up navigable rivers in search of loot, land, and women. Wars have been won and lost by sea power or the lack of it. Coastal blockades and the occupation of strategic straits and islands have been used as methods to assert sea power. Gradually a body of international law has developed, supplemented by many bilateral and multilateral treaties, concerning access to the sea, navigation of the high seas, and the rights and obligations of those who use the high seas. Salvage rights, safety standards, the right to seek shelter in time of storm, questions of signaling and communications between ships at sea and between ship and shore have been worked out in some detail. The administrative and judicial rights of sea captains have been standardized. The rights of belligerent and non-belligerent vessels in time of war have been proclaimed if not always observed. Exclusive fishing zones, policing zones, and pollution control zones have been announced by individual nations, and some measure of international understanding has been reached in these matters.

The maritime nations of the world reached agreement at the recent Law of the Sea conferences in Geneva on the way to define the outer limits of the land by drawing a base line along the shore and across bays and estuaries. They failed to agree, however, on a standard width for territorial waters. The exploitation of minerals on and under the continental shelf has raised new questions of definition and delimitation. Some useful agreements have been reached concerning the international allocation of mineral rights in certain areas of great international interest such as the Persian Gulf and the North Sea.[3] As the technology of mineral production from the continental shelf and from sea water has progressed, new questions have been raised concerning ownership of the sea-bed resources which might be found under what is now international water. Having in mind the enormous wealth that American and international oil companies are winning and planning to win from the continental shelves, the representative of Malta at the United Nations General Assembly in 1967 proposed that "the sea-bed

[2]Frank Quinn, "Water Transfers: Must the American West Be Won Again?", *Geographical Review*, vol. 58, January, 1968, pp. 108-132.

[3]A. Denis Clift, "North Sea Gas: A Case Study in International Cooperation," *The World Today*, vol. 23, April, 1967, pp. 146-152.

and the ocean floor are a common heritage of mankind and should be used and exploited for peaceful purposes and for the exclusive benefit of mankind as a whole." The Malta Declaration, which is being given serious consideration, went on to declare that the needs of poor countries should receive preferential consideration in the event that exploitation of the ocean floor becomes commercially profitable.[4]

With the advent of atomic energy and atomic and hydrogen bomb testing, questions have been raised concerning the pollution of the international atmosphere. Although the pollution of international water by the dumping of waste materials is already the subject of a number of international agreements, it appears that these are inadequate for handling some of the recent problems. The disposal of atomic wastes, the accidental dumping of whole cargoes of oil from large tankers like the *Torrey Canyon* off the English coast, and the damage done by the detergents used in the clean-up operation following such tanker accidents all call for closer international regulation and supervision. The recent Santa Barbara oil leak during the process of offshore drilling along the California coast was another stern reminder of the dangers of water pollution in the oceans.[5]

Large-scale international exchange of information about water resources and water technology is an encouraging development. Two major international conferences held in Washington, D.C., within the past five years are worth noting.

The First International Symposium on Water Desalination was held on October 3-9, 1965. Invitations were sent to 130 nations and, of these, 54 sent official delegations. The papers presented at this conference, touching on almost every imaginable aspect of water desalination, have been published in three large volumes totaling more than 2,250 pages.[6] Part of President Johnson's address, which was read to the assembled delegates by Donald F. Horning, Special Assistant to the President for Science, is of interest here:

> Techniques to desalt water have been used in many places for many years—on ships at sea, among the islands of the Caribbean, in desert lands along the Persian Gulf. But if our own vision for the future is to be realized—the vision of an inexhaustible supply of pure, drinkable water—then the cost of desalting must be drastically reduced. Research results have been heartening. We have built five plants capable of testing new

---

[4]Arvid Pardo, "Who Will Control the Seabed?", *Foreign Affairs,* vol. 47, October, 1968, pp. 123-137.

[5]Abel Wolman, "Pollution as an International Issue," *Foreign Affairs,* vol. 47, October, 1968, pp. 164-175.

[6]*Proceedings of the 1st International Symposium on Water Desalination, Washington, D.C., October 3-9, 1965,* Washington: Government Printing Office, 1967. 3 vols.

technologies. Their daily capacities range from a few hundred thousand gallons to more than two million gallons. We have built and operated more than a score of pilot plants. We have witnessed the cost of desalted water cut in half and then halved again. To accelerate this work we have recently launched a new five-year, $200 million program of research and development. We have concrete goals in view: by 1968, to construct plants with the capacity of 10 million gallons per day; by 1970, to extend the range to 100-million gallon plants. We are also at work on smaller plants varying in size from less than one million gallons to 15 million gallons and employing many different processes. From the creative work you perform in your laboratories and on your drawing boards, and from your conferences like this one, we will gain new freedom from the harsh accidents of geography.[7]

Scientists and engineers from many different countries have contributed to the international effort looking toward more efficient water desalination. During the 1965 conference, reports were presented on a great variety of processes: the use of solar stills, freezing techniques,

FIGURE 8.1 Making Fresh Water from the Ocean. Courtesy U.S. Office of Saline Water. The world's largest experimental desalting unit is a multi-stage flash distillation module recently completed at the San Diego Test Facility of the Office of Saline Water. It furnishes 2.5 million gallons per day of fresh water to the city of San Diego, and is a prototype of much larger units to be constructed later.

[7] *Ibid.*, vol. I, pp. 5-6.

ion-exchange methods, biological desalination, multi-effect-multi-stage flash distillation, osmotic desalination with porous membranes, and others. The stakes in this research are tremendous, for the world's population is increasing by more than one and a half percent per year, and the people's demand for fresh water is increasing much faster.

Even larger than the First International Symposium on Water Desalination was the International Conference on Water for Peace, held in Washington, D.C., May 23-31, 1967. With 94 country delegations and 24 international organizations participating, it was one of the largest international meetings ever held in the capital city. More than 600 prepared papers, numerous speeches, and 52 country-situation reports were presented, all of them published in the eight-volume conference report.[8] Once more President Johnson's words were heard by the assembled body of engineers, social scientists, politicians, and others in the audience. These are again worth quoting:

> Last week I signed an Act of Congress to make possible a new plant which will more than double the world's present capacity for desalting water. A decade ago, the best plant design could produce only 50,000 gallons per day at a cost of $5.00 per 1,000 gallons. . . . This new plant, powered by nuclear energy, will eventually produce 150 million gallons of fresh water per day at a cost approaching 20 cents per 1,000 gallons. This is three thousand times as much, at 1/25th the cost. But the world needs water at even lower costs.[9]

The new desalination plant to which President Johnson was referring was the one being built on the coast of California at San Diego as a joint project of the Department of Interior, the Atomic Energy Commission, and the Metropolitan Water District of California. The whole world is hoping that it will be as successful as the designers predict. Water is a universal substance in universal demand, and new techniques of producing and managing it, wherever developed, are as welcome in Saudi Arabia and Algeria as they are in China, Connecticut, and California.

---

[8] *Water for Peace, International Conference, Washington, D.C., May 23-31, 1967*, Washington: Government Printing Office, 1968. 8 vols.

[9] *Ibid.*, vol. I, pp. 10-11.

Index